Riparian Landscapes examines the ecological systems of streamside and floodplain areas from the perspective of landscape ecology. The specific spatial pattern of riparian vegetation is seen as a result of, and a control on, the ecological, geomorphological, and hydrological processes that operate along rivers.

The landscape structures of world riparian zones, seen as their longitudinal and transverse spatial configurations, are compared. Internal structures are seen as the configuration of tesserae representing spatially disjunct environmental gradients.

Riparian structures are controlled by the spatial dynamics of channels, flooding and soil moisture, and human impact. These dynamics are part of integrated cascades of water, sediment, nutrients and carbon, with the riparian zone acting as source, filter, and sink, to which animal and plant species respond through dispersal and invasion in ways that illuminate diversity, community structure and competition. The role of the riparian zone in controlling species distribution and abundance is discussed. Intelligent management of these valuable ecological resources is highlighted. The potential for linking hydrological, geomorphological and ecological simulation models is also explored.

This book will be of interest to graduate and professional research workers in environmental science, ecology, forestry and physical geography.

Riparian landscapes

Cambridge Studies in Ecology presents balanced, comprehensive, up-to-date, and critical reviews of selected topics within ecology, both botanical and zoological. The Series is aimed at advanced final-year undergraduates, graduate students, researchers, and university teachers, as well as ecologists in industry and government research.

It encompasses a wide range of approaches and spatial, temporal, and taxonomic scales in ecology, experimental, behavioural and evolutionary studies. The emphasis throughout is on ecology related to the real world of plants and animals in the field rather than on purely theoretical abstractions and mathematical models. Some books in the Series attempt to challenge existing ecological paradigms and present new concepts, empirical or theoretical models, and testable hypotheses. Others attempt to explore new approaches and present syntheses on topics of considerable importance ecologically which cut across the conventional but artificial boundaries within the science of ecology.

CAMBRIDGE STUDIES IN ECOLOGY

ALSO IN THE SERIES

H. G. Gauch, Jr — *Multivariate Analysis in Community Ecology*
R. H. Peters — *The Ecological Implications of Body Size*
C. S. Reynolds — *The Ecology of Freshwater Phytoplankton*
K. A. Kershaw — *Physiological Ecology of Lichens*
R. P. McIntosh — *The Background of Ecology: Concept and Theory*
A. J. Beattie — *The Evolutionary Ecology of Ant–Plant Mutualisms*
F. I. Woodward — *Climate and Plant Distribution*
J. J. Burdon — *Diseases and Plant Population Biology*
J. I. Sprent — *The Ecology of the Nitrogen Cycle*
N. G. Hairston, Sr — *Community Ecology and Salamander Guilds*
H. Stolp — *Microbial Ecology: Organisms, Habitats and Activities*
R. N. Owen-Smith — *Megaherbivores: the Influence of Large Body Size on Ecology*
J. A. Wiens — *The Ecology of Bird Communities:*
Volume 1 Foundations and Patterns
Volume 2 Processes and Variations
N. G. Hairston, Sr — *Ecological Experiments*
R. Hengeveld — *Dynamic Biogeography*
C. Little — *Little Terrestrial Invasion: an Ecophysiological Approach to the Origins of Land Animals*
P. Adam — *Saltmarsh Ecology*
M. F. Allen — *The Ecology of Mycorrhizae*
D. J. Von Willert, B. M. Eller, M. J. A. Werger, E. Brinckmann, H. D. Ihlenfeldt — *Life Strategies of Succulents in Deserts*
J. A. Matthews — *The Ecology of Recently-deglaciated Terrain*
E. A. Johnson — *Fire and Vegetation Dynamics*

Riparian landscapes

GEORGE P. MALANSON

Published by the Press Syndicate of the University of Cambridge
The Pitt Building, Trumpington Street, Cambridge CB2 1RP
40 West 20th Street, New York, NY 10011-4211, USA
10 Stamford Road, Oakleigh, Melbourne 3166, Australia

First published 1993

Printed in Great Britain at the University Press, Cambridge

A catalogue record for this book is available from the British Library

Library of Congress cataloguing in publication data

Malanson, George Patrick, 1950–
Riparian landscapes/George P. Malanson.
p. cm. – (Cambridge studies in ecology)
Includes bibliographical references (p.) and index.
ISBN 0-521-38431-1 (hardback)
1. Landscape ecology. 2. Riparian ecology. I. Title. II. Series.
QH541.15.L35M35 1993
574.5′222 – dc20 92-30617 CIP

ISBN 0 521 38431 1 hardback

SE

Contents

Preface

We were curious. Our curiosity was not limited, but was as wide and horizonless as that of Darwin or Agassiz or Linnaeus or Pliny. We wanted to see everything our eyes would accommodate, to think what we could, and, out of our seeing and thinking, to build some kind of structure in modeled imitation of the observed reality. *(John Steinbeck)*

I shudder to begin a book with this quote. I do not have any real hopes of meeting Steinbeck's goals nor, certainly, of matching his prose. It is important to note, however, that I am trying to give some kind of structure, albeit personal, based on observed reality, but for the most part on reality at second hand. I have worked in riparian landscapes in Iowa, Montana, and Utah, but in this book I primarily review the fairly large collection of scientific literature on riparian ecology from the perspective of landscape ecology. Certainly more than Newton, I stand on the shoulders of giants, and I have tried, perhaps to the extent of overzealousness, to document their work in detail. The work is diverse, addressing themes from pollination biology to sediment deposition. I believe that landscape ecology can build a useful structure in modeled imitation of the observed reality because the various processes share a common spatial structure and real location. In the process of constructing this structure I hope also to provide some insights into just how useful a blueprint is landscape ecology for such construction. I have chosen it as a starting point because I believe that geography, including both relative space and specific space, is important.

Acknowledgements

While not specific to this work, my association with Walt Westman as a student and as a colleague has been educational, daunting, and inspiring. I owe a debt of gratitude for the hours of discussion and fieldwork with graduate students, present and past, at the University of Iowa: David Bennett, Mary Craig, Sandra Finley, Jeff Hanson, John Kupfer, Christopher Lant, Zhi-Jun Liu, K.D. Rex, Bill Schwarz, and Louise Zipp, and with many undergraduates, including Dan Ray, John Drennan, and Ed Nealson. My friend and colleague David Butler has provided useful comments on the manuscript, as well as participating in some of my riparian (and other) research. Richard T.T. Forman encouraged me along the way and made valuable suggestions. Conversations with Martin Hermy at Herent, Jean-Paul Bravard and his colleagues at Lyon, Henri Decamps and his colleagues at Toulouse, Guy Pautou and his colleagues at Grenoble, and Ferdinand Vasicek and his colleagues at Brno helped to shape this work, and I appreciate their hospitality. My colleagues at the University of Iowa, especially Graham Tobin, Richard Baker, Frank Weirich, Marc Armstrong, and Rebecca Roberts have contributed. Not least, I am grateful to Mary McCoy Malanson for bearing with me.

This work began under funding from the National Science Foundation, grant SES-8721868. The work of writing was made possible by a Faculty Scholar Award at the University of Iowa.

1 · *Principles considered*

> The Mississippi, the Ganges, and the Nile, those journeying atoms from the Rocky Mountains, the Himmaleh, and Mountains of the Moon, have a kind of personal importance in the annals of the world ... Rivers must have been the guides which conducted the footsteps of the first travellers ... They are the natural highways of all nations, not only levelling the ground and removing obstacles from the path of the traveller, quenching his thirst and bearing him on their bosoms, but conducting him through the most interesting scenery, the most populous portions of the globe, and where the animal and vegetable kingdoms attain their greatest perfection. *(Henry David Thoreau)*

Recent work in international ecology has focused attention on an area called 'Landscape Ecology'. Several recent symposia and books have pointed out a connection between the interaction of biotic and abiotic structures and functions and their spatial organization (Zonneveld 1990). In this book I will use some of the paradigms of landscape ecology to organize knowledge about riparian environments, and I will use that knowledge to assess those paradigms. While much of my approach will be toward ecological problems, I will also consider problems of a more broadly defined physical geography which are held in common with landscape ecology.

Disciplines

Landscape ecology

Landscape ecology has arisen from practical considerations of how ecological ideas could be applied in land management. The idea developed in Europe, where land management, including nature management, is intensive, and where the land has been clearly divided for centuries. Schreiber (1990) reviewed the development in Europe, giving much of the credit to Troll (e.g. 1939, 1968), and credits other German physical geographers in the 1960s (e.g. Neef 1963; Schmithusen 1963). Landscape ecology was introduced to North America in the 1980s (Forman and Godron 1981; cf. Forman 1990). In 1986, an American chapter of the International Association for Landscape Ecology was formed at the International Congress of Ecology. R.T.T. Forman pioneered this introduction with others (e.g. Risser *et al.* 1984; Naveh and Lieberman 1984), building on a sound reputation in ecosystems studies which he had applied to a whole region in New Jersey (Forman 1979). Forman and Godron's (1986) landmark work found a ready audience.

One reason for the ready audience in North America has not been given the attention that I believe that it deserves: the legacy of R.H. MacArthur. MacArthur (1972; MacArthur and Wilson 1967) developed the idea, in island biogeography, that space mattered. In island biogeography, distances, locations, stepping-stones, shapes, and areas have an effect on the sizes, dynamics, and organization of populations and communities. This work recognized that all habitats were islands of some degree and that spatial considerations applied in all places. These ideas were followed upon by many studies of isolated places. As a geographer trained in a spatial tradition (e.g. Abler *et al.* 1971) and intrigued by patterns on maps, I found that these studies served as a basis for a focus on spatial pattern and process. These studies often found that the simple relationships of area and isolation with immigration and extinction rates were complicated by various other environmental and spatial factors. In particular, early efforts to use the principles of island biogeography to address the design of nature reserves were quickly seen to have limitations (e.g. Simberloff and Abele 1976). These applied problems for real places demanded a more complex approach to the spatial aspects of ecology.

Landscape ecology is that approach, because it provides a framework and a sense of direction for the unformed ideas on how and why space mattered. Traditional biogeography sought to explain spatial patterns, with some reference to spatial processes. It did not, however, consider the dialectical nature of spatial pattern and process. Core concepts of landscape ecology, such as the mosaic controlling the flux of energy, matter, and information, are new developments that distinguish it from biogeography.

Physical geography

In essence, landscape ecology in the broad sense is the mostly abandoned core of physical geography, i.e. the spatially mediated interaction of environmental processes that creates the distributions of climates, life, and landforms on earth. Orme (1980), in reviewing the status of physical geography as part of a series published in *The Professional Geographer*, noted that in the early twentieth century physical geography consisted of efforts to classify and synthesize environmental processes of places or regions, but that in recent decades physical geography had been seen as an introduction to advanced study in the more specialized fields of climatology, geomorphology, etc. Physical geographers have been waiting for a unifying theme. Although the systems paradigm as presented by Chorley and Kennedy (1971) was meant to unify divergent interests in physical geography, the development of detailed systems analyses or systems related studies within separate areas became the dominant method of research.

Orme (1980) argued that in order for physical geography to be viable it should be definable. He wrote:

> But the subtle combination of subject matter and methodology does seem to create, in a societal context, a unique and definable role for physical geography, namely as a spatial and temporal explanation of natural phenomena at or near the earth's surface, with particular emphasis on the interrelationships among phenomena and between these and society.

The other three papers in the series in *The Professional Geographer* were specifically on biogeography (or ecological geography) (Vale and Parker 1980), climatology (Mather *et al.* 1980), and geomorphology (Graf *et al.* 1980). These works each stressed a unique role that physical geographers could make in their respective areas. Two themes are common: a spatial approach, expressed as environmental systems with topological and topographic relationships; and the interaction of people and environment. These two themes are held in common with landscape ecology. Because their research is better appreciated and recognized in allied disciplines rather than in geography itself, physical geographers had become increasingly specialized in these three areas, and the relationships among them had been disregarded. The central theme had been lost.

Landscape ecology, in part because it arrived in America with a developed tradition and recognition, has been readily accepted as a unifying theme by ecological geographers. In a discussion of the relationships among the three fields at a panel session during the Association of American Geographers annual meeting in 1987, the distance was obvious. The suggestion that landscape ecology may provide a way to rediscover the lost core was not met with enthusiasm by those in climatology or geomorphology, where it has yet to make an impact. Landscape ecology remains an ecological discipline. Landscape ecologists will continue to benefit from understanding the effects of the geomorphological development of a landscape and the climatological processes at landscape scale on the biota, but the flow of information in the other direction will take time. Geomorphologists and climatologists are beginning to acknowledge the importance of the biota for their concerns: two books, titled *Biogeomorphology* (Viles 1988) and *Vegetation and Erosion* (Thornes 1990), show the influence, and climatologists are beginning to recognize that the regional feedbacks from the biota and soil are critical to understanding regional and global climate. The development of geographic information systems has provided a tool by which data and models in the three areas that include spatial variation can be reconciled. This avenue for the future will be considered in the context of riparian environments.

The role of history

Debates have arisen in ecology over the successes and potential of reductionistic and holistic approaches to the field. No immediate resolution is in sight. It is worthwhile, however, to consider some aspects of our ontology and epistemology as we consider landscapes. In geography, this debate was once considered as a dichotomy between idiographic and nomothetic study. An idiographic approach considered every place to be unique and the description of this uniqueness to be a goal of study (e.g. Hartshorne 1939). The nomothetic approach, as the word implies, sought general laws that explained processes in many places (e.g. Harvey 1969). In geography this dichotomy was voiced by those who upheld the old tradition of regional geography and those who, in the 1950s and 1960s, led the quantitative revolution into thematic geography and especially spatial analysis. Physical geography, while moving directly into the nomothetic path, became so involved with process that the importance of either spatial relationships or place was absent from most actual research although acknowledged in theory (cf. Gregory 1985).

In ecology the distinction between idiographic and nomothetic approaches has been blurred, and in fact confounded. A major distinction has been made between what have been characterized as mere description and functional study. In fact it has been the population and community ecologists concerned with the processes of evolution who have worked in the nomothetic tradition by deriving hypotheses from theory and testing them with observations about the structure of populations and communities. Systems ecologists, on the other hand, have produced elaborate descriptions of the functions of ecosystems, but they have not, until recently, engaged in nomothetic science. Although their descriptions shared methodology and vocabulary, they did not propose or test theory.

Landscape ecology today draws from both of these traditions in the philosophy of science. Much of the emphasis has been on the nomothetic tradition. Forman and Godron (1986) cited many of the works regarded in geography as its foundations as a spatial science. But landscapes are in a sense unique. No two landscapes are the same, so that replication of units for study is impossible. Moreover, the role of history in the development of landscape pattern and process is overwhelming; no two landscapes have had the same history, and much of the history is unknowable. Decamps and Fortune (1991), while explaining the needs for sensible long-term research designs for riparian landscapes, emphasized that knowledge of the history of the landscape is critical to understanding its processes. In this situation, it is necessary to reevaluate the feasibility of a single or single-minded attempt at understanding. More recently in geography the debate between idiographic and nomothetic approaches has been reformulated and restated as a debate on the importance of place and space in the context of a contrast between logical positivism (sometimes as a strawman) or its replacement, realism or critical rationalism, and historical materialism. Suffice for now to say that at larger spatial scales a holistic approach seems needed and history limits the application of replicated reductionist methods.

Landscape reproduction

Landscape ecology is based on the hypothesis that the interactions among biotic and abiotic components of the landscape are spatially mediated. Not only are the flows of energy, material or species from place to place affected by the locations of the places in the landscape, but these flows then determine the interactions among energy, material and species. In ecosystem models, the processing of energy by species, for example, depends on the population size and available materials (e.g. nutrients for metabolism). Landscape ecology identifies how the availability of energy, material, and populations will be affected by their location, but exactly how these ideas might be operationalized, and hypotheses derived and tested, in real landscapes has not been explained.

My interpretation of the central theme of landscape ecology is that spatial structure controls the processes that continuously reproduce that structure. This continuous reproduction can be expressed as a dialectical development, and in some cases the difference between the process and the structure will need to be defined carefully. This concept of landscape reproduction needs to be considered in terms of actual rates of change and differences in the time scales of changes. This dynamic landscape development can be seen in terms of the feedbacks among processes which are affected by their locations or configurations in space. Not only the overall heterogeneity of landscape elements, but also their specific topology, affects the flows of nutrients, energy, and species on the landscape.

The dialectical concept implies a continuous relationship, while feedbacks imply discrete time steps. This difference is in part due to the modeling techniques applied in ecology wherein discrete time intervals often are specified for given processes. The best example of this is in the introduction of generation times in population models. Continuous models using differential equations in fact behave differently from discrete models using difference equations. In a landscape, some processes will operate at distinct annual cycles, while others will be continuous. Some may even have longer discrete periods. None the less, we can consider that landscapes are reproduced. While it may be too glib to use an organismic analogy, we could consider that as landscapes reproduce, they also evolve as the reproductive process includes variations in spatial pattern that are either more or less successful in a given environment.

When feedbacks are thought to operate, lag times must also be considered. Lag times in the effect of a feedback mechanism lead to nonequilibrium conditions (Malanson *et al.* 1992). Disequilibrium may be a fundamental component of ecological systems (DeAngelis and Waterhouse 1987). Landscape ecology is a particularly appropriate approach to the study of nonequilibrium systems because it takes into account the relations among the several processes of physical geography, and because nonequilibrium environments have distinct relations between hydrology, geomorphology, and ecology (Butler and Malanson 1990; Malanson and Butler 1990; Kupfer and Malanson 1992a). Moreover, landscape ecology focuses on one of the specific components of time lags: spatial separation.

Riparian landscapes for landscape ecology

Forman and Godron (1986) spent considerable effort on defining terms in landscape ecology. While some (e.g. isodiametric, i.e. circular), are unwieldy, most are useful. Most of the terms that they used are standard English and are used in ways that are not contradictory. In an area that is new, or at least unfamiliar to many, beginning with a common vocabulary is helpful, and Forman and Godron (1986) have served us well in this regard. These general definitions will be used here, and their definitions repeated if necessary.

A central problem for landscape ecology has been the definition of the smallest unit of study that makes up a landscape (as individuals clearly make up a population). Forman and Godron (1986) identified landscapes as made up of 'landscape elements' which in turn are made up of tesserae. They noted that how finely an area is divided, and thus what are the most homogeneous units and what groups they form, are not exact and may best be communicated by example. They used the example of an agricultural area. Elements are fields, farmyards, woods, and roads. Tesserae are the smallest homogeneous units visible, e.g. individual corn fields or woodlots. For riparian areas, the example is not so clear because of the continuity of the river, and in some cases the riparian vegetation, over long distances. In general, landscape elements in the riparian zone will include distinct vegetation types, wetlands, and other land-use categories. Tesserae, which I will discuss in terms of the internal structure of riparian zones, would be levees, abandoned channels, and their less distinct forms as ridges and swales within a floodplain forest. While differences may exist in the identification of elements in a real landscape, the concept is clear enough for the general discussion which follows. The title of this book, *Riparian Landscapes*, should not be interpreted to mean that a narrow river corridor is a landscape in and of itself, but rather that the riparian zone is a functionally dominant feature which contains and connects elements.

In a landscape which is made up of a number of ecosystems, the flows of energy, matter, and species are determined to some extent by the spatial configuration of the elements. Forman and Godron (1986) identified seven principles of landscape ecology which directly address these relationships. Being hypotheses or principles under development, some of them seem tautological, depending upon and deriving from the definition of landscape elements. The general idea is that landscapes are made up of several elements, and that the degree of heterogeneity, and thus the size of elements and edges, affects the interaction, or flows of energy, material, and species, among the elements. The heterogeneity also affects the way in which landscapes respond to disturbance through affecting flows and thus resistance and recovery.

The seven principles are:

Landscapes differ structurally in the distribution of species, energy and materials, and therefore differ functionally in the flows of species, energy, and materials among the elements.

Landscape heterogeneity decreases interiors, increases edges, and enhances species richness.

The changes in the distributions of species are controlled by landscape heterogeneity, which is in part defined by these distributions.

Nutrient flows in the landscape increase with disturbance.

Flows of energy and biomass across boundaries increases with heterogeneity (i.e. the number of boundaries).

The intermediate disturbance hypothesis applies to landscape heterogeneity as well as to species diversity.

Landscapes will develop either physical system stability, resilience, or resistance to disturbance.

While ecosystem studies have focused on such flows and interactions, they have missed the importance of spatial organization because they have looked within landscape elements or tesserae rather than among them. Even within landscape elements the importance of spatial location has often been generalized. Huston *et al.* (1988) have shown that identification of the location of individual trees in a small forest area will produce different results in a model of forest dynamics in comparison with spatial averaging, but most models are spatially averaged. Ecosystem concepts have treated the location of a given ecosystem in a landscape as its boundary conditions, and have proceeded to model the interactions with the inputs from and outputs to surrounding landscape elements as given. This approach, as pointed out by many workers in landscape ecology and biogeography, limits the explanatory power of the ecosystem paradigm. In two other fields closely associated with riparian studies, hydrology and geomorphology, spatially explicit models are more common.

Forman and Godron (1981, 1986) defined three major aspects of landscape structure: patches, corridors, and matrix. Patches are distinct landscape elements surrounded by others. Patches may be distinguished by their origin, size, and shape. Corridors are narrow landscape elements, differing from the surrounding elements. Like patches, they may be distinguished by origin, and size takes on characteristics of width, and shape takes on characteristics of sinuosity. The matrix is simply the dominant element in a landscape, and Forman and Godron (1986) give three criteria for distinguishing a matrix, which are inverse in their order of importance relative to their ease of definition: the largest area, the greatest connectivity, and the most control over dynamics. The overall pattern of patches and corridors within a matrix of a landscape may include networks in which patch number and configuration, corridor connectivity, breaks and nodes, boundary shapes, and overall heterogeneity operate to affect the flows of energy, matter, and species and their interactions. These ideas will be examined relative to riparian landscapes.

Riparian environments have received considerable attention from both ecologists and fluvial geomorphologists, but this work has not been given a unified conceptual framework. Here I define *riparian* in the broader ecological sense of the word, rather than the more restrictive sense of within the actual banks of the river (as preferred by C.R. Hupp, personal communication). In the broader sense, *riparian* includes the ecosystems adjacent to the river (this usage is long standing, as evidenced by the *Oxford English Dictionary*'s example for *riparial*). To try to use *floodplain* would be misleading because the riparian zone includes narrow strips along downcutting rivers, islands, and channel landforms as well as extensive floodplains. From the perspective of the ecologist, the dynamic processes of erosion, deposition, and water flow are considered as impacts upon the biota, and little attention as been paid to the reverse effects and essentially none to how the location of the ecosystem under study is important. Fluvial geomorphologists have considered the feedback between vegetation development and the processes of water and sediment movement, but here too the studies are specific to individual sites and little locational context has been considered. Decamps and Naiman (1989) presented an outline of the landscape level concerns for riparian ecology, and Naiman and Decamps (1990) presented a group of papers that go far in addressing some of these concerns, but feedbacks and the effect of space on the expression of environment in place still need investigation. In that book, Risser (1990) identified the importance of riparian ecotones for consideration of key current environmental issues, and noted that they can be a useful locus for testing ecological ideas. Gregory *et al.* (1991) have also noted the unique and important role of riparian ecosystems in a landscape setting. They specifically cited the linear spatial configuration and its role in increasing the interaction of the riparian zone with surrounding ecosystems.

Landscape ecology can provide a unifying concept to the diverse interests in the structure and dynamics of plant and animal communities, the trapping of sediments and nutrients eroded from agricultural fields, the alteration of flood hydrographs, and the development of landforms. Riparian environments provide a place where the hypotheses of landscape ecology can be operationalized and tested. Whereas Forman and Godron (1986) provided a framework for describing and classifying the spatial relations in a landscape, in the riparian environment the spatial relations are clarified. Their concept of corridor is critical here as a starting point (cf. Forman 1983). They note that a number of measures might be applied to a corridor: breaks or connectivity, variations in width, nodes or intersections, all of which combine to determine another of their points, the network. Because of breaks in the corridor, ideas applying to patches also must be considered: their size, shape, number, and configuration. The origin of the corridor landscape elements is also important. Landscape ecology also focuses on how the pattens affect the processes, particularly in relation to the control of flows of energy, matter, and species among landscape elements. Such flows are integral; the major ones well known in riparian areas, and the spatial structure of riparian areas, will need to be considered in respect of their functions as conduits and as barriers to these flows.

The aim of this book is to conceptualize the diverse work done on riparian environments, particularly in plant ecology, but extending into geomorphology, hydrology, and agricultural economics, in terms of landscape ecology and physical geography. Much of Chapters 4, 5, and 6 are in the nature of a review from this viewpoint. The interactions among the ecological, geomorphological, and hydrological factors that make up the riparian landscape are highlighted by examining how computer simulation models can be linked together using geographic information systems in order to provide a basis for testing hypotheses about the role of space. Specific dynamic models that emphasize the temporal aspects of one factor in one place are presented as components of a framework that allows the study of the interaction of components through time in diverse spatial conditions. Different spatial scales of focus are discussed. The implications of this approach for landscape ecology, for physical geography, and for the practical application of these concepts to land management are considered.

Conclusions

Landscape ecology is an approach to the study of the environment that emphasizes complex spatial relations. The relative locations of phenomena, their overall arrangement in a mosaic, and the types of boundaries between them, become the priorities of study. Spatial arrangements are not necessarily reducible to general rules, however: the history of individual places makes each unique. A landscape is continuously reproduced, as processes create patterns which in turn control the processes. Riparian environments are well suited for the elucidation of principles of landscape ecology: their ecology has been studied, their spatial characteristics are relatively clear, and they are found everywhere. This book explores how riparian environments can be seen in terms of landscape ecology.

2 · *Riparian topics*

Some Plains Indians, like the Hidatsa, believed shadows cast by cottonwoods possessed intelligence and would counsel a troubled person. While all 'the standing peoples' have voice, few are so sweetly loquacious as the gentle and generous cottonwood. *(William Least Heat-Moon)*

The riparian landscape is unique among environments because it is a terrestrial habitat strongly affecting and affected by aquatic environments; it has a particular spatial configuration; it has use values derived from these features; and, like mountain or desert habitats, is diverse in its structure and function among regions while responding to the same primary factors. This work will primarily address the effect of the particular spatial configuration of riparian environments, as they differ among regions, on both the structures and functions in the environment.

Basic definitions

The link between the terrestrial and the aquatic realms in riparian zones is the most significant factor in their definition, although Forman and Godron (1986) prefer to view them as a corridor rather than an ecotone. Brown *et al.* (1979) began with a definition that might encompass most wetlands, i.e. 'a high water table because of proximity to an aquatic ecosystem or subsurface water' and 'an ecotone between aquatic and upland ecosystems'. They then qualified this definition to include 'only those types which are exposed to lateral water flow... We assume that lateral water flow becomes the main force that organizes and regulates the function of riparian wetlands including their biogeochemical cycles and their role in the landscape'. Graf (1985) simply defines riparian as 'in and near river channels and directly influenced by river-related processes'. In this book the importance of the lateral surface flows of water will be investigated (Figure 2.1). While other flows are also important, and the subsurface lateral flow of water must be distinguished, it is this particular surface flow that distinguishes riparian habitats from others. A

Figure 2.1 Lateral water flow through riparian vegetation during over-bank flooding, St. Francis River, Missouri.

number of structural and functional concerns are introduced when one concentrates on the form and processes of a river, and the traditional ecological concept of 'riparian' has meant affected by the river. In this work deltas will be considered only to a limited extent because the riverine processes are greatly modified by the lentic or marine system.

Brinson (1990) described riparian forested wetlands in some detail. His review shares some concepts with this work. Most notably, he considered the biogeographic location of sites, but only reported on the floristic differences. He did not go into any detail on the geographic differences among processes, and differences in landscape pattern are mentioned only in passing. In a major section aimed at biomass and productivity, Brinson (1990) examined many of the processes that will be discussed here. In particular he examined aspects of the hydroperiod in this regard. I too will examine the hydroperiod, but primarily in reference to the community structure and species dynamics of riparian sites. By way of contrast Brinson (1990) provides a view of another approach to this topic.

Terrestrial-aquatic gradient

Ecotones have received increasing attention in recent years. Holland (1988) reported on a SCOPE/MAB workshop on the topic. Hansen *et al.*

(1988) argued that ecotones at a variety of spatial scales could be considered. They considered three aspects of ecotones that are central questions in landscape ecology: flows of energy, matter, and organisms; biodiversity related to spatial location; and landscape management. They also noted that ecotones, because they represent the range limits (at least on a local environmental gradient) for species, may be sensitive to regional or global climatic change. Naiman *et al.* (1988*b*) reported on a specific UNESCO program on land–inland water ecotones, many of which will be riparian zones. It will need to be recognized that riparian areas may be an ecotone at one scale, but may themselves have two ecotones, the upland and the aquatic, at another scale. Much of the emphasis has been on the terrestrial–aquatic ecotone. Risser (1990) noted the importance of this distinction for examining the concepts of landscape ecology, specifically how a sharp environmental gradient would affect biodiversity and flows of matter and energy.

The ecotonal character of riparian habitats is variable. In some places the gradient may be obvious because it is short, as opposed to an ecocline where the same change in the general environment is spread over a greater distance. Conversely, in some areas the riparian zone may be so distinct that it appears to be totally isolated from the upland. The rate of change in ecosystem in these landscapes is determined in large part by the relation between the topography and the hydrological regime of the river. One aspect that will be considered is the historical legacy of nonequilibrium conditions between the present topographic form and the climogeomorphic regime that created them.

The wetland character of riparian habitats is also variable. Some areas that are immediately adjacent to rivers are not wetlands. They may not be flooded frequently and, depending on channel incision, may not have a high water table. Here the soil may seldom be saturated and the vegetation may be the same as that of upland sites. From the more xeric areas, a gradient exists to true aquatic conditions. Some riparian areas contain lakes, ponds, and wetlands which are perennially inundated. These aquatic habitats are beyond the scope of detailed examination of this book, but their presence as landscape elements or tesserae will be considered. Along this gradient are frequently inundated wetlands, vegetation occasionally flooded, and areas infrequently flooded. The riparian environment will be examined in relation to the hydrological regime in order to examine this gradient.

Riparian values

Riparian landscapes also concern land managers and social scientists because they are affected by water resource developments and associated land use. This investigation will concentrate on the interaction of ecological and physical processes, but these are not completely divorced from human activities. Riparian vegetation plays a role relative to erosion, channel stability, and water quality, and riparian areas have aesthetic, recreational, and resource values more directly related to humans (e.g. Brown and Daniel 1991). A variety of papers presented at symposia have addressed the economic, social, and legal aspects of riparian lands (e.g. Meyer 1984, 1985; Wakeman and Fong 1984). Brinson *et al.* (1981) tabulated some values of riparian ecosystems. Taking this as a starting point, I list potential values in Table 2.1. Brinson *et al.* (1981) and Desaigues (1990) also discussed in some detail the methodology, merits and problems of assigning value to natural lands and processes.

Riparian areas have aesthetic and recreational values (Figure 2.2). Yon and Tendron (1981) identified them as attractive landscapes near the population centers of Europe. Even in these long-settled areas they are seen as 'untamed' and symbolizing '... a nature almost untouched by man'. Lant (Lant and Tobin 1989, Lant and Roberts 1990) has also identified the aesthetic value of riparian areas, and with it their recreational value. Cole and Marion (1988) likewise focused on recreation but examined its impacts in riparian areas. Hunt (1988) detailed the wildlife values of riparian lands while examining catastrophic losses due to federal water projects in the USA. Riparian values may be even more important in less developed countries, but identification and quantification of values in some areas is difficult (Madgwick 1988).

An indirect economic benefit of riparian areas is the maintenance of water quality as an environmental service (Roberts and Lant 1988, Lant and Roberts 1990); this process is discussed in detail in Chapter 5. Graf (1978, 1980, 1982, 1985) has discussed the relation between riparian vegetation and channel stability for the Colorado River. While detailing processes that will be discussed in this book, he has also noted their importance for the overall management of this river basin, which is one of the most intensively used on earth. Graf (1985) also notes the chemical instability in the river basin, a topic that has received considerable attention for rivers in agricultural regions. Proposals have been made for using riparian vegetation as filters for agricultural nutrients as well as

Table 2.1 *Values of riparian ecosystems*

Extant riparian systems:
Economic
Reduce downstream flooding
Recharge aquifers
Surface water supply in arid regions
Support secondary productivity, e.g. for fisheries
High yields of timber
Social
Recycle nutrients, tighten spiral and storage
Store heavy metals and toxins
Accumulate organic matter as a sink for CO_2
Intermediate storage for sediments
Natural heritage
Recreation
Aesthetics
Natural laboratories for teaching and research
Biological
Special habitat for some endangered or threatened species
Refugia for upland species
Corridors for species movement
Former riparian systems
Economic
Transport corridors
Water supply and electricity
Construction materials and waste disposal
Agriculture and livestock
Settlement

traps for agricultural sediment (see Chapter 5). Recent US agricultural policy has designated riparian zones as eligible for inclusion in the US Conservation Reserve Program, in which farmers can be paid for not farming certain environmentally sensitive areas (Cohen *et al.* 1991). Riparian areas also contribute to the quality of river fisheries (e.g. Barton *et al.* 1985; Crance and Ischinger 1989; Platts and Nelson 1989; Leitman *et al.* 1991).

Direct economic benefits, often in competition with benefits of channel stability or aesthetics and recreation, may also accrue from riparian lands. Two notable ones are from livestock grazing and forest

Figure 2.2 Recreational and aesthetic values of riparian areas include sports activities.

harvest (Boldt *et al.* 1979). Zube and Simcox (1987) have investigated some conflicts related to grazing riparian vegetation in arid areas (Figure 2.3). Knopf (1988), Sedgwick and Knopf (1987, 1991) and Knopf *et al.* (1988) have examined the relation between grazing, vegetation, and animal populations in riparian areas in Colorado. They found that grazing did affect the horizontal pattern of vegetation and thus animal populations, but not in a harmful way if managed correctly. Leonard (1988) examined how managers could effectively regulate grazing in public lands. The US Forest Service has extensively considered forestry practices in relation to river channels, and now prohibits clearcutting adjacent to streams (Anderson 1985; Steinblums and Leven 1985). In some areas, the riparian zone may be the only feasible source of wood or forage. Athearn (1988) noted that, in the western USA, riparian areas were the first to be used by Euro-Americans (in travel) and the first to be settled for ranching and farming. Some of the earliest economic activity by Euro-Americans in the western USA, beaver trapping, exploited the riparian landscape and had far-reaching consequences for it (e.g. Parker *et al.* 1985). In less-developed countries, riparian habitats may be a critical part of the resource base (Douthwaite 1987). Lawton (1967) noted the importance of riparian forest sources for timber in Zambia, and Ansari (1961) proposed that riparian forests be used for timber production

Figure 2.3 Riparian bottomland in the western United States has been developed extensively for grazing; Virgin River, Utah.

rather than fuelwood in Pakistan. The complexity of such resource questions has led to unresolved conflicts and new attempts at resolution (Pinay 1988; Pinay *et al.* 1988).

Other potential values require the destruction of some riparian areas (Table 2.1). Losses of riparian habitat are difficult to estimate but are considered to be great. Yon and Tendron (1981) noted that such destruction began centuries ago in Europe, but that riparian forests did continue to exist. Wenger *et al.* (1990) more closely documented losses in Europe, but noted that complete information was scarce. Brinson *et al.* (1981) estimated that 70% of the natural riparian communities in the USA had been lost. They reported losses as high as 96% and 98% for

Table 2.2 *Forces leading to the destruction of riparian habitat*

Direct
Agriculture and livestock
Mining
Industry
Transportation and communication
Urbanization
Indirect
Dams
Canals
Dikes
Pollution

southeastern Missouri and for the Sacramento River, California, respectively. In the southeastern USA, Turner *et al.* (1981) presented a variety of estimates of riparian loss for hardwood forests. They reported a 56% reduction in area since 1937, at which time some areas had already been converted, mostly to agriculture, following the Swamp Lands Acts of 1849–50. Louisiana suffered a loss of 50% (a change in riparian forest cover from 42% to 21% of the total area of the state). Figures for other southern states are variable and may reflect differences in the surveys. They noted that these riparian hardwood forests recovered some area in a period from 1940 to 1960, but have since resumed a decline. These changes in trend result in a complex landscape. In California, Katibah (1984) estimated that 11.1% of original riparian forest remained in the Central Valley, of which 5.3% was degraded and the remaining 5.8% was in the process of being degraded.

The exemplary catalog of the forces of destruction presented by Yon and Tendron (1981) can be used to examine this process in general (Table 2.2). They categorized destruction as either direct or indirect, and as directed toward either core or fringe habitats. The most notable direct forces are those of agriculture (Figure 2.4); industry, mining, and urbanization are secondary direct effects. Indirect effects are primarily those that alter the hydrological regime, i.e. dams, canals, dikes, etc. (Figure 2.5), and secondary processes are those of uses for forestry and recreation and also pollution. The types of impacts on riparian areas vary by region. For example, Turner *et al.* (1981) show projections of a conversion of 9.5 million acres (3.8 Mha) of riparian forest to cropland in

Figure 2.4 Riparian areas have been cleared of natural vegetation for agricultural development in many places; White Breast River, Iowa.

the Mississippi River alluvial floodplain for 1950 to 2000. In the western USA, however, many of the impacts are directly or indirectly the result of dams; streambank 'protection' is also a factor (Shields 1991). In Europe, riparian areas have been affected by a combination of these forces (Wenger *et al.* 1990). Girel and Manneville (1991) have shown, however, that complex and valuable ecosystems can be maintained with some necessary human impact, such as the mowing of floodplain fens. Where the ecosystem has developed in conjunction with a specific and limited human impact, the challenge may be to maintain the specific practice if it becomes economically unviable. Although in some areas the type of impact is identifiable, less distinct impacts may also be important: Mason and MacDonald (1990) compared river reaches in more and less developed areas of Great Britain. They found distinct differences in the number of trees per kilometer of river and documented the rate of decline in the developed area from 1879 to 1970.

The direct effects on riparian areas, where the natural landscape is simply replaced by other conditions, are easy to see but still require a sociological explanation in order to be understood fully. Hunt (1988) gave several examples of direct effects. She reported that along the southern portion of the Mississippi River, in the broad floodplain, transformation of hardwood forests to soybean fields continued in spite

Figure 2.5 Riparian forest land lost to reservoir flooding, Iowa River, Iowa.

of legal efforts to preserve wetlands. She documented how federal policies in fact promoted this conversion. She also documented another direct impact, the Tennessee–Tombigbee waterway, in which a canal for shipping destroyed considerable riparian habitat, and yet carried only 6% of the traffic projected by its promoters. This example is but one of several where the economic returns of the transportation system do not pay for the costs of the project, even given the undervaluation of natural lands. In Europe, such projects have been in place for centuries, and while the costs and benefits are impossible to calculate, the loss of riparian habitat is certain (Bravard 1987). Another direct effect on riparian habitats comes from their flooding when reservoirs are constructed. Fleshman and Kaufman (1984) reported the decline in riparian habitat following temporary flooding for a single year in the flood pool of a California reservoir. This loss is total and has been documented in many places, and again Hunt (1988) has presented case studies that apply. For one of these, Shanks (1974) reported that the Oahe Dam in South Dakota flooded 160 000 acres (64 750 ha) of bottomlands on two Native reservations. Hunt (1988) noted that this project was one of several planned to provide flood protection for White towns, none of which were displaced.

Reservoirs are also one of the primary factors in the indirect impacts on riparian habitats. Below the reservoir, the river flow regime is altered.

The variations in flow are lessened. Flood peaks and low flow periods are both reduced. This change in flow pattern can have a major impact on riparian areas. Impacts of a dam on the riparian ecosystem, particularly on grasses, ungulates, and waterbirds, of the Kafue and Zambezi Rivers in Zambia have been reported (Sheppe 1985; Pinay 1988). Hunt (1988) again provided several examples of such impacts. For example, the Colorado River in the southwestern USA provides several case studies of the effects of the changed regime. In Europe, Bravard (1987) provided a detailed analysis of the indirect effects of a variety of water use programs on the riparian environment of the upper Rhone. On both rivers we see the destruction of riparian vegetation by longer periods of inundation at moderate levels, and the lack of regeneration owing to more stable substrates in the absence of high magnitude flooding. Mining also has an indirect impact on riparian systems. Gravel mining from the bed of rivers in France has had a notable impact on the channel dynamics of alluvial rivers such as the Rhone, and thus on riparian vegetation (J.P. Bravard, personal communication). Specific ecological effects of such impacts will be discussed in Chapter 4. Indirect effects may not always be so obvious, as in the case of general plans for regional development (Hughes 1984).

Reisner (1986) provided a detailed look at the political motivation in the USA for the development of water projects that have adverse environmental effects, and Worster (1985) added a theoretical explanation for this type of development in a capitalist society. Much of the destruction stems from the failure to recognize the value of natural riparian habitats in contrast to the direct economic benefits of water and land development projects. This failure is not only due to the difficulty of quantifying such benefits, but also because the monetary benefits of development projects accrue to the wealthy with political influence, while the benefits of nature are diffuse. Hunt (1988) reported repeated instances where the US Army Corps of Engineers, responsible for many development projects, ignored the judgements of the US Fish and Wildlife Service, responsible for protecting biota. Clark (1980) and Knopf *et al.* (1988) noted differences between US agencies and their role in conserving riparian landscapes,which arise from different historical, philosophical, and legal missions.

One problem with the assessment of changes in the riparian landscape is knowledge of previous, or in some cases present, conditions (Abernethy and Turner 1987). Turner *et al.* (1981) presented a map of US wetlands adapted from Kuchler (1964); this map includes no riparian wetland vegetation along rivers in most of the upper Mississippi River valley, nor along any rivers in the eastern USA north of Washington, DC. Lugo *et al.* (1990) include a global map of forested wetlands, and the category of 'Alluvial' includes only the Mississippi embayment in North America, no areas whatever in Europe, Asia, or Australia, and only some scattered sites, seemingly at the headwaters of the Zaire River, in Africa. Areas in South America seem better covered. These omissions may be a matter of the map scale and definition, but they indicate the problem for assessment. Satellite-based mapping systems, such as used by Hewitt (1990) for inventory of riparian ecosystems in Washington, provide a means to solve this problem, but new and better sensors will be needed for the effort.

Riparian values are diverse, and although each one may not affect many people, in sum these values have implications for future river corridor use. Diverse and widespread values mean that effective political coalitions can be formed if compromises between competing objectives can be found and if an area of common good can be identified (cf. Gruntfest 1991).

Landscape perspectives

The general spatial patterns of riparian areas can be conceived in the terms of landscape ecology proposed by Forman and Godron (1981, 1986). The spatial configuration of riparian areas are those of corridors, but the features of patches also apply. Moreover, they form a network within an overall matrix. At this scale, riparian areas have features in common, and these commonalities, and the contrasts between them for riparian areas in different regions or affected by different forces, will be instructive. The landscape pattern of riparian areas is controlled by the drainage patterns that are characteristic of geological and climatic areas. Many drainage patterns are a variation of dendritic, with differences in the angles of branching and the parallelism of the channels. Other factors that will affect the overall pattern are the history of the region, especially in the Holocene, and the more recent human impacts. These general considerations will be described here, and their actual differences examined below in examination of the landscape and internal structures of riparian areas in different regions.

Spatial structure

As corridors within landscapes, riparian environments have special spatial configurations. The physical structure of a corridor is clear. The riparian corridor within the matrix may or may not be distinct; Forman and Godron (1986) place some emphasis on the visual interpretation of landscape elements, but the superficial visual characteristics as well as the underlying factors may vary in how distinct a distribution they have on the landscape. In some areas it may be that the riparian zone appears to be the matrix, as defined by Forman and Godron (1986). Riparian elements may be the most extensive, depending on where the boundaries of the region are drawn; the riparian area may often be the most connected, by virtue of the continuity of the river; and the riparian zone may exert more control on the flows in the landscape, especially on the movement of water, nutrients, sediment, and species.

Genesis

Riparian areas are essentially environmental resource patches. It is the nature of the hydrological regime of the place that distinguishes them from their surroundings. In some areas the resource is the abundance of water, which allows the development of broadleaved plants; in others, it is a limitation on productivity, specifically caused by anaerobic conditions. These aspects of landscape control will be examined among regions and also in terms of system functions. In the general terminology of Grime (1979) (where disturbance is the destruction of extant growth by outside factors, e.g. fire or herbivory; stress is the reduction in rate of growth by outside forces causing a shortage of resources, e.g. cold or dry climates; and competition is, in the absence of the other two, limitations on growth imposed by a shortage of resources due to capture by other individual plants) this represents a gradient from stress to competitive conditions that is based on the productivity of the site, and in landscape terms we can consider the productivity of the site relative to surrounding landscape elements. Simple productivity is not a sufficient measure, however, because of differences in the factors limiting productivity in different places in the landscape.

This primary resource patch factor may be combined with two others. Riparian landscape elements may also be due to disturbance: while riverine processes alter resource conditions, they may also have direct density-independent effects on the mortality of individuals. Also, riparian areas may have features of remnant or regenerated elements where human land use has intervened. The shift of rivers across alluvial plains, the active erosion at river banks, and the continual entrainment and deposition of sediment up to boulder size are aspects of hydrological disturbances that destroy biota. The floods that riparian areas experience

Figure 2.6 Once continuous riparian forest cover fragmented by agriculture and other land uses, Iowa River, Iowa.

may occasionally be disturbances, but are more often stress. Notable disturbances are those caused by people. The scope of such disturbance has been discussed above. Where land use, especially deforestation and drainage, have altered riparian areas, the natural riparian area may be in the form of remnant patches. In some areas previously used and abandoned areas may have the characteristics of regeneration patches. For example, along the western tributaries of the Mississippi River which extend into grasslands in the central USA, what was once more continuous riparian corridor is now divided into a string of patches (Figure 2.6). These natural forest remnants can be compared to an archipelago of forest in a sea of corn and beans.

Network

Riparian areas may also be easily examined as a network. The usual dendritic pattern presents one distinctive network, and the drainage density and degree of integration of drainage patterns will provide different network structures. While the rivers are usually continuous, the riparian corridor may not be. Breaks in forest cover caused by people is one example of a discontinuity. Such breaks will have different consequences for biotic and abiotic flows. Hanson *et al.* (1990) showed, for example, that dispersal of riparian forest species by squirrels might be limited by large agricultural fields. The sinuosity of the river will also have an effect on continuity, because major meanders may bring riparian areas close together that are on separate sections of the river. The effects of patches of riparian vegetation on opposite sides of the river and the extent to which the river itself operates as a barrier are additional considerations.

Shape

The shape of the riparian zone is also important. The shape of the network, discussed above, is that created by the resource. Modification of this basic pattern by disturbances, creating disturbance, regeneration, and remnant patches, make shape of further interest. Shape is important because it determines the relation between area and perimeter. This distinction is important because of what is known in ecology as an edge effect: the species composition and abundance found in the edge area is different from that in the interior of a landscape element. Species can sometimes be classified as either edge or interior species. These differences between edge and interior arise because of specific climatological effects, responding to basic processes in environmental physics. Forman and Godron (1986) presented several means for defining shape parameters and identified gradients in processes that would be associated with changes in interior to edge ratios. In addition to completely filled patch shapes, rings are a possibility, as would be found around the shore of a wetland, and peninsulas present another particular shape. Rex and Malanson (1990) identified three major factors affecting the shape of remnant riparian forest patches in Iowa (Figure 2.7). Human land use was most important, resulting in straight-lined edges of patches. River sinuosity was important, because meander bends are so pronounced that some patches closely approximate circles with smooth edges. The width of the river valley was also important because in wide valleys more complex shapes may develop. We noted that useful shape indices could be important as independent variables in future exploratory studies.

Width

The width of a riparian corridor may have other important consequences. Forman and Godron (1986) emphasized this aspect especially in relation to how a corridor would act as a conduit of species movement on the landscape. This width is important for the interaction of the riparian zone as a whole with both the surrounding matrix and with the river. In this way the width will affect the heterogeneity of the riparian

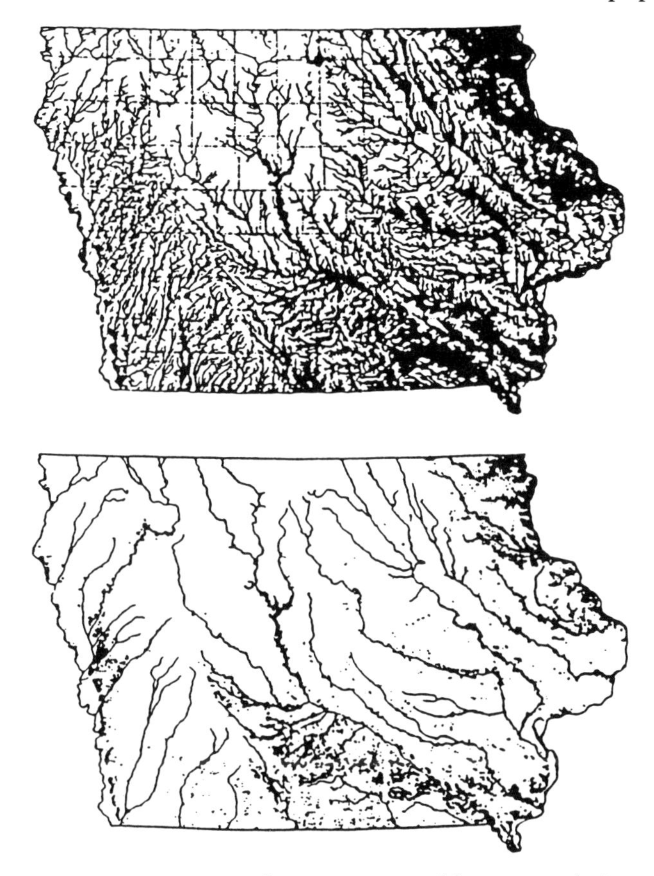

Figure 2.7 Former and present extent of forest cover in Iowa. In pre-settlement times most forest cover was riparian, except in the northeast. Now, upland woodlots have expanded through planting and fire protection, while riparian lands have been developed. The balance is obvious.

element, i.e. the number and kinds of tesserae that develop within it. In general, wider riparian areas will be found on the wide floodplains of alluvial rivers and heterogeneity will be increased. On such floodplains the extent of the riparian areas may be so great as to reduce the relative importance of the connectivity with uplands in their function as a conduit. The width may be even more important when the riparian zone is considered as a barrier, either to material flux from uplands to the river or to species movement across the landscape.

Configuration or pattern

Spatial features are quite variable among riparian landscapes. Baker (1988) presented four ways in which the spatial structure of riparian

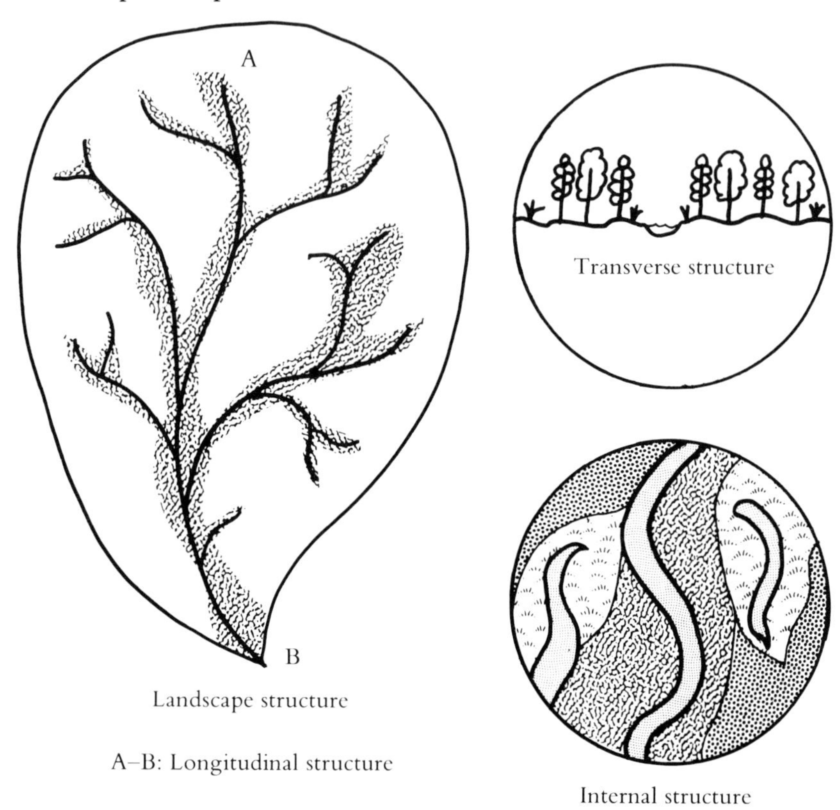

Figure 2.8 The scale of investigation of landscape and internal structure in riparian environments, and the difference between the longitudinal and transverse landscape structures.

vegetation can be considered: within-stand variation, between-stand variation, within-region variation, and between-region variation. Here I will examine pattern within regions, but differences among them should also produce insights. I will consider two major divisions of pattern: landscape structure and internal structure (Figure 2.8).

Landscape structure includes the pattern of the riparian environment within the surrounding matrix. Two aspects will be emphasized: the longitudinal structure from the headwaters to the mouth of rivers (cf. Naiman *et al.* 1987 for an in-stream consideration of this gradient), and the pattern of the riparian zone within basins, or transverse structure, which may be considered as a cross section of the floodplain (Amoros *et al.* 1987*b* also discuss a vertical dimension in another context) (Figure 2.8). Hupp (1990) presented a discussion that explicitly considers both gradients for riparian vegetation. The boundaries of the riparian zone with the surrounding matrix and its own connectivity will be examined. Particular cases will focus on regional differences: boreal and subalpine forest, deciduous hardwood forests, forest–prairie transition and beyond, and scrub deserts. Although the driving forces are the same, differences in the strength of the forces, and thus in the general landscape, have resulted in significant differences among riparian environments in different regions. These differences are informative in respect to the relative importance of various processes in determining the structure of plant communities in the riparian zone.

The patterning of plant communities and their relation to landforms and hydrological processes within the floodplain will be examined as the internal structure of riparian elements (the arrangement of tesserae within the element). Most research has focused on this area, and has emphasized a simple gradient away from the river. The complexity of the floodplain will be presented by reference to studies of processes of hydrology and fluvial geomorphology, and the response of plant communities will be examined relative to soil moisture and anoxia, flooding, and fluvial dynamics. The gradients within local areas are determined by these factors, but other factors, such as fire, grazing, and permafrost, may alter the basic relations. These gradients can be generalized as a transverse gradient perpendicular to the channel, but actual locations, especially along alluvial rivers, are more complex. Topographic variation is the common element. A gradient of microclimate, in part related to light penetration and soil moisture, probably exists within riparian zones but has not been quantified; it too may have some relation with topography. These abiotic gradients, through their influence on plant physiological and morphological characteristics, control the distribution and abundance of plant species and the diversity of communities.

Processes

The flows or processes by which riparian landscapes change will be examined in detail. Aspects of material, energy and species dynamics will be considered. The processes by which water, sediment, nutrients, and contaminants interact will be presented, with special attention to the relation between water quality and agricultural landscapes. The interaction of riparian areas with the hydrological regime of rivers will be examined in terms of the erosion, transport, and deposition of sediment which defines the topography, which in turn is the primary control of the interaction. A consideration of the linkages between these processes and vegetation dynamics will highlight the long-term processes of the riparian zone.

Ecological processes

The functions of riparian elements can be conceived in systems terminology. They form the inputs, regulators, stores, throughput, and output of cascading systems. As such they operate at intersections with the structural elements to reproduce new structure through time. The functions of the riparian zone operate on energy, matter, and information (where species are equivalent to genetic information) cascades. Ecosystem research has focused on energy and especially nutrient transport, but studies have concentrated on one place, and the spatial structure has been ignored. Hydrological research has more completely included spatial considerations in distributed parameter, spatial explicit models of drainage basins which do model the cascades of water and sediment. The studies in ecology that link together aspects of these functions either do so in a limited area, without a systems approach, or consider a limited number of factors.

The function of a corridor is more complex. For animals that use the corridor for migration the function is obvious: it serves as a pathway. For tree species that may use the corridor for diffusion in response to climatic change over the course of centuries, the function is more obscure. Some of the functions of the riparian landscape elements will operate relative to their position between the upland and the river, i.e. as a barrier rather than as a conduit: riparian zones can filter or stop movement. The flows of sediment, water and nutrients in the riparian zone are affected by both the transverse pattern and width of the element, and by its longitudinal pattern. The riparian area has aspects of both conduit and barrier in this sense. Direct transfers of energy will also be affected by both aspects of the pattern. The longitudinal pattern primarily has an effect in the downstream direction. The more complex spatial configuration summarized by the two major gradients will of course complicate the functions.

Relatively little is known about specifically riparian processes related to microclimates, photosynthesis, and the production and transport of litter, and the gross importance of woody debris has been identified but not tied into other system functions. Studies of riverine ecology have identified relations between inputs of carbon and patterns of abundance of organisms in rivers in a paradigm called the river continuum concept (Vannote *et al.* 1980), and the distribution of these energy transformation processes in relation to the longitudinal and transverse pattern of the riparian landscape will be considered in a parallel way with specific reference to the nutrient spiraling concept (Newbold *et al.* 1981). The spatial and temporal fluxes in the riparian zone are not continuous, however, and these concepts will need modification (Junk *et al.* 1989).

Vegetation processes, following from the specific discussion of the vegetation structure of the riparian landscape, will be considered as flows of genetic information. The influence of the landscape and internal structures on ecological dynamics will be examined. The spatial configuration of riparian areas as corridors, the directional force of the flow, and the dispersal pathways of the species will be discussed. The processes of regeneration will be emphasized, but the interactions of regeneration, growth, and mortality will be linked to abiotic and interspecific processes.

Geomorphological processes

Because riparian ecology is related to the processes of the river, the geomorphic setting is important (cf. Marston 1991). Rivers flow in diverse areas, and consequently fluvial geomorphology is a broad topic. Several key points will be worth examining, however. Two aspects of river morphology and two of fluvial landforms should be considered. River morphology can be considered in terms of channel form and channel pattern; fluvial landforms are either erosional or depositional, and in most areas the two coexist.

The characteristics that are defined for channel form are width, depth, slope, roughness, and associated velocity and discharge. These characteristics derive from the composition of the bed and bank materials and flow regimes, which in turn depend on regional climate and geology. Morisawa (1985) discussed the factors determining the hydraulic geometry of channels. For riparian landscapes, it is important to note that rivers with low sediment loads have more ability to entrain and carry additional sediment than those already near 'capacity', and that channels in easily eroded material will tend to widen, while those in more resistant material, except where the bottom is specifically armored, will tend to incise, or deepen. Morisawa (1985) described five channel patterns, which can be put into two groups: with or without multiple channels. In single channel rivers the pattern is a function of sinuosity, and rivers can be either straight, sinuous, or meandering. Multiple channel rivers are

(*a*)

Figure 2.9 (*a*) Extensive braiding in a wide valley of a montane river carrying large loads of coarse sediment. (*b*) A relatively stable mid-channel island; Middle Fork of the Flathead River, Montana.

either braided, with unstable channel bars, or anastomosed, with stable islands (Figure 2.9). There are perhaps some overlaps especially in large alluvial meandering rivers where the process of meandering itself can create secondary channels and a river appear anastomosed. Morisawa (1985) lists both erosive and depositional behavior associated with the five types.

The regional erosive feature of interest is the stream network. Small streams merge to form large rivers. This process is formalized in a descriptive method called stream order (Morisawa 1985) in which stream order increases from 1 at headwaters to larger numbers depending on the number and sequence of junctions. Stream networks can take on a variety of patterns, depending on the geology, geomorphology, and climate.

Other erosive features of interest include incision. Canyons and gullies are features that can spatially isolate riparian zones (Figure 2.10). These features are common in areas of geological uplift, such as in mountain ranges, or on a smaller scale, where the base level has dropped. The canyons of the Colorado Plateau present a spectacular and interesting

(b)

example of uplift and incision. The ancestral Colorado and its tributaries meandered widely. Uplift resulting from tectonic forces allowed the rivers to incise while maintaining their meandering pattern. In the area of Cataract Canyon on the Colorado River, removal of overlying material reduced pressure on underlying deposits of potash, which expanded and created an anticline (Molenaar 1985). This canyon now runs along the axis of this breached anticline, providing a textbook example of inverted topography. While these incised meanders reveal the geological history of the area, they have resulted in unique spatial distributions of riparian ecosystems.

The process of increased sinuosity, i.e. meandering, is important. Because of the hydraulics at the bend of rivers, meanders migrate. Erosion at the outside of the bend means that meanders migrate outward

Figure 2.10 Deep canyons can isolate riparian elements from the surrounding landscape; Cataract Canyon, Colorado River, Utah.

and downstream. The lateral erosion of any river, and especially of meanders, creates cutbanks, which are the most distinct edges of the riparian zone. The degree of transition from the terrestrial to the aquatic environment in the riparian ecotone can be drastically altered by this effect (Figure 2.11). Meanders are also associated with deposition. Point bars are deposited at the inside of a meander bend (Nanson 1980) (Figure 2.12). Because the average channel form does not change greatly, point bars follow the meanders in their migration, thus always providing new areas of deposition. Meanders do not move in a regular fashion, and so the radius of curvature changes; an upstream meander, moving fast, can cut off the next downstream one, and the river will abandon that channel to take the shortest course. In combination, these processes mean that the floodplain will be topographically and sedimentologically complex. Braided and anastomosed channels occur where banks are erodible, gradients are steep, bed load is abundant, and/or the river lacks the competence to carry sediment. Which factors are causal is not clear. Braiding does seem to occur where large amounts of coarse bed load can be mobilized by temporary high discharges which then lose competence. The result is deposition of bars in mid-stream (Figure 2.9*a*). These bars, and the islands of anastomosed rivers, can present unique spatial patterns of riparian habitat.

Figure 2.11 A steep and high cutbank in a terrace of paraglacial gravels isolates the vegetation from the riverine processes; Middle Fork of the Flathead River, Montana.

Floodplain deposition followed by renewed downcutting can leave terraces of depositional material, but these have also been termed erosional features, depending on details of the process. Over millennia a sequence of such terraces can develop, presenting major topographical features within the river valley, but the riparian zone, affected by the current river flow, is within the most recent terraces.

Non-equilibrium conditions

The overall pattern of a riparian area within its region may be in response to forces that are no longer operating, and the pattern may be in transition toward an equilibrium with current forces. Climatic and geomorphic changes are globally ubiquitous, and thus so are nonequilibrium riparian environments.

The degree of disequilibrium varies by region. The most notable nonequilibrium areas are those described as paraglacial (Church and Ryder 1972). In these areas glaciation in the Pleistocene produced massive amounts of sediment, which still fills local valleys in the mountains and covers extensive areas of continents. These sedimentary deposits still condition the riparian landscape. Another related aspect is that of underfit rivers. In some places rivers draining the melting glaciers

Figure 2.12 Deposition at the inside of a meander bend leads to the development of point bars, which are important sites for riparian vegetation; Middle Fork of the Flathead River, Montana.

carved valleys that are much larger than would be created by the present flows (Salisbury *et al.* 1968). These relatively large valleys also condition riparian landscape processes. Along the upper Mississippi River and in some sections of its tributary streams, such as in Iowa, the floodplains are more than 2 km wide and have steep bluffs that separate them from the uplands. The topographic distinction of these trapezoidal notches in the landscape affects present ecological processes although created by Pleistocene hydrology.

Nonequilibrium conditions also arise from historical and ongoing human impacts. Such conditions are now considered to be fundamental in ecological systems, although the degree of disequilibrium may be a function of the spatial scale of the investigation (DeAngelis and Waterhouse 1987). In riparian landscapes the spatially transgressive nature of fluvial processes leads to nonequilibrium conditions through the repeated creation and destruction of different tesserae. The continual erosion of cutbanks and creation of point bars in alluvial rivers are examples of ongoing geomorphological processes that may be a delayed response to hydrological and land use changes, and which are in turn responded to by changes in vegetation and wildlife. The nature and degree of disequilibrium and its effect on riparian landscapes will be considered throughout this book. The implications of disequilibrium for landscape ecology and landscape management are important. In at least one sense, disequilibrium can lead to deterministic chaos (Malanson *et al.* 1992).

Conclusions

Riparian areas are unique environments because of their position in the landscape. They are both ecotones between the terrestrial and aquatic zones and corridors across regions. Riparian environments serve diverse functions and have different values depending on their physical, biological, and cultural setting. The difference in values has often led to the destruction of some values for the sake of others. From the perspective of landscape ecology, riparian areas have qualities that make them well suited for study, and a landscape perspective may also help to judge the alternative values of these places.

3 · *Landscape structure*

Going up that river was like travelling back to the earliest beginnings of the world, when vegetation rioted on the earth and the big trees were kings. *(Joseph Conrad)*

Having defined the landscape structure of riparian environments as the spatial pattern of the riparian zone within its region, including the major gradient from the source to the mouth of the river and its width, more detail is needed on how landscape structure can be operationalized or used to generate hypotheses about process. Forman and Godron (1986) considered several concepts that address the spatial configuration of landscape elements within the landscape. In reference to riparian landscapes, I will examine the concepts that pertain specifically to corridors, such as curvilinearity or sinuosity, whether the corridor is higher or lower than its surroundings, and connectivity, and others that apply to patches and the matrix, such as size, shape, and boundary characteristics, which are also important in the riparian context. One approach is to compare the landscape structure of different ecoregions.

Ecoregions

Bailey (1983, p. 365) defined ecoregions as 'geographical zones that represent geographical groups or associations of similarly functioning ecosystems'. Bailey (1976) produced a map of ecoregions in the USA, which Teskey and Hinckley (1977*a*,*b*, 1978*a*,*b*,*c*) later used to classify riparian ecosystems. Bailey (1980) often noted the structure of the riparian vegetation in his description of individual ecoregion provinces in the USA. Although Omernik (1987) produced an ecoregion map for the USA which takes land uses into account (the Iowa River area changes from Bailey's Humid Temperate Domain, Prairie Division, to Omernik's Central Corn Belt ecoregion) and which has been used for classifying in-stream ecological conditions (Rohm *et al.* 1987), Bailey (1989) has more recently produced an ecoregions map of the world, which includes four Domains divided into 31 Divisions, 29 of which are further divided into 96 provinces. I will refer to this conceptualization, but I will not produce a region-by-region description. Among riparian areas, some greater similarities are found between rather than within domains.

A very important part of the landscape structure of riparian areas is created by the fact that the rivers flow across more than one ecoregion. The most persistent such gradient is where rivers flow from highlands to lowlands (mountain areas are given a separate ecoregion designation at the second degree division by Bailey (1989)). The gradient from wet to dry regions is sometimes associated, although most rivers flow from continental toward maritime climatic regimes (exceptions occur where rives flow into desert areas, e.g. the Okavango delta in southern Africa). Rivers also flow from warm to cold regimes and vice versa. The fact that several ecoregions are linked is one facet of interest in riparian landscape ecology.

Here, I will examine seven types of river landscapes in related ecoregions (Table 3.1): arid and semi-arid gallery forest landscapes (based primarily on the southwestern USA in the Dry Domain, Subtropical Desert Division and the Humid Temperate Domain, Mediterranean Division, typified by the Colorado River or the San Joaquin and Sacramento Rivers, and in Africa in the Savanna Division, typified by the Niger and Pongolo Rivers); tropical forest landscapes (primarily in the Amazon basin, in the Humid Tropical Domain, Rainforest Division); subtropical floodplain forest landscapes (primarily in the southeastern USA in the Humid Temperate Domain, Subtropical Division, typified by the Mississippi River embayment); humid broadleaf forest landscapes (primarily in the northeastern and north–central USA and western Europe in the Humid Temperate Domain, Warm and Hot Continental Divisions, and the Marine Division, typified by the Rhone and Ohio Rivers); forest–grassland transition and grassland landscapes (primarily in the Midwest of the USA, in the Humid Temperate Domain, Prairie Division, and the Dry Domain, Temperate Steppe Division, typified by the Missouri River); mountainous landscapes (primarily in the western USA, in both the Subtropical Steppe Regime Division typified by the Flathead River, Montana, USA, and the Marine Regime Mountains Division, typified by the Willamette River, Oregon, USA); and taiga landscapes (primarily in Canada and Sweden in the Subarctic Division, typified by the Mackenzie River, Canada). While I have identified particular rivers here, I will also refer to

Table 3.1 *Environments for which riparian landscapes are described*

Arid and semi-arid gallery forest (Subtropical Desert Division, Mediterranean Division, and Savanna Division)
Tropical forests (Rainforest Division)
Subtropical floodplain forests (Humid Temperate Domain, Subtropical Division)
Humid broadleaf forests (Humid Temperate Domain, Warm and Hot Continental Divisions and Marine Division)
Forest-grassland transition and grasslands (Humid Temperate Domain, Prairie Division and Dry domain, Temperate Steppe Division)
Mountains (Dry Domain, Temperate Steppe Regime Mountains and Humid Temperate Domain, Warm Continental Regime Mountains and Marine Regime Mountains)
Taiga and tundra (Subarctic and Tundra Divisions)

similar rivers or analogous situations from the same domain, division, or province. Although the bulk of the available research is from sites in the USA, I have tried to include as much work from other areas as possible. I will then consider linkages among regions in summary.

Arid and semi-arid gallery forest (Subtropical Desert Division, Mediterranean Division, and Savanna Division)

In terms of general vegetation, Mediterranean-type, desert and semi-arid scrub, and savanna plant formations are very distinct. I have grouped them together for the discussion of their riparian component because they share some common landscape features. These three regions share in common some period when the climate is dominated by subtropical high pressure. The desert regions are more or less continually dominated; the Mediterranean-type regions are dry in summer, and the savannas in winter. These seasonal droughts lead to overall lower productivity, lower biomass and cover, and the consequent climogeo-

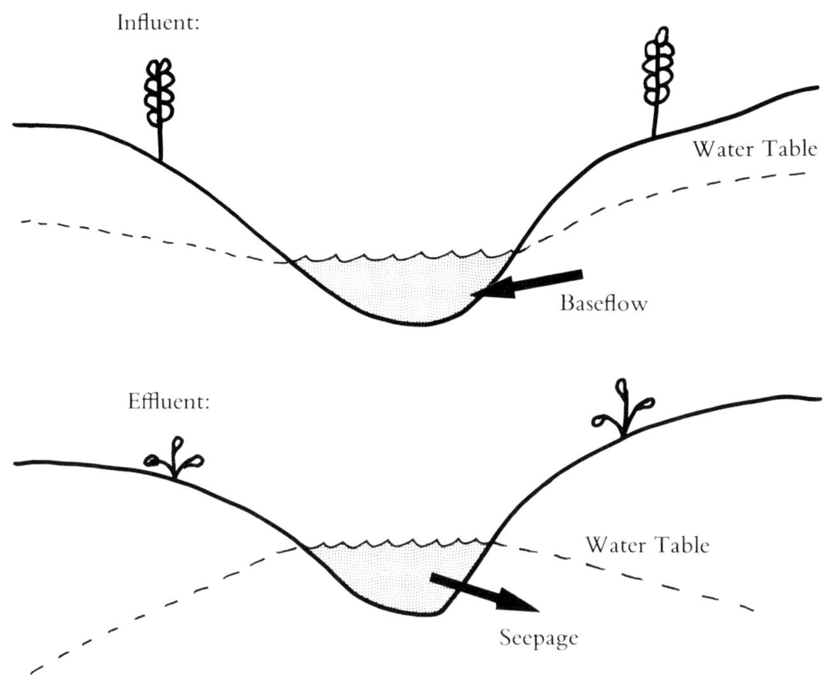

Figure 3.1 In influent conditions , groundwater moves into the channel and maintains baseflow. In effluent conditions, water moves from the channel into a lower groundwater region; this later state can be maintained only by exotic rivers, i.e. those with sources in regions with influent conditions.

morphic conditions that set the stage for their riparian landscape ecology. The rivers in these areas are often exotic, i.e. their flow is maintained by precipitation in source regions in mountains, and their relation with groundwater is influent, i.e. water moves from the surface flow of the river into the ground, rather than vice versa (Figure 3.1). These conditions create particular spatial constraints on the biota. Another important aspect of riparian areas in dry environments is that the streams are often intermittent and/or ephemeral; i.e. they are spatially and/or temporally discontinuous. Intermittent streams are almost always also ephemeral, but many streams can be ephemeral, i.e. seasonal but spatially continuous. Courtois (1984) described an extreme case in which riparian ecosystems become established only in occasional years of high precipitation and are replaced by halophytes in dry years.

Most notable is that these corridors are extremely visually distinct because they are higher than the surrounding vegetation (Figure 3.2)

Figure 3.2 In arid areas the vegetation of the riparian element is more productive and stands higher than the surrounding landscape.

(some grasslands are discussed separately because of the nature of ecotones and geomorphology), and/or are deciduous elements in an evergreen landscape (Holstein 1984). Forman and Godron (1986) contrasted corridors lower and higher than the surrounding vegetation, but in fact they discussed only the effects of corridor width. Relative height is important because it affects microclimate and may affect flows of energy, material and species to and from surrounding elements. Height also may indicate productivity. The productivity of riparian vegetation in these environments is lower than in the riparian environments of other regions, but it is higher than that of the surrounding matrix of shrublands, desert scrub, or savanna grasslands. The limiting resource in these regions most often being water, the rivers are the sites of the highest potential production and thus the largest accumulations of biomass. The distinction between the riparian vegetation and that of the surrounding matrix is, however, often very sharp, because the depth to the water table increases greatly with distance from the river; this feature is characteristic of exotic streams. Thus the defining characteristics of riparian landscape structure in these areas are its height above the matrix and the sharp ecotone.

Another feature of these areas is that their rivers are often incised into

Figure 3.3 In the southwestern USA, and in other arid and semi-arid areas, where the river has incised into the landscape the vegetation of the riparian landscape is isolated from the surroundings; Colorado River, Utah.

canyons, further confining the vegetation and modifying the landscape structure of the riparian zone (Figure 3.3). The canyons of the southwestern USA present notable examples of clear instances where riparian vegetation is confined to canyon bottoms. Uplift of relatively flat-lying sedimentary formations has been accompanied by the downcutting of preexisting rivers, producing spectacular canyons including deeply incised meanders. In such cases, the landscape ecology may be simplified. In studies of hanging gardens (which are not, however, everywhere confined to riparian locations) in the bottom of the Narrows of the Virgin River in Utah, I was able to examine the effects of relative location by assuming that the canyon bottom represented a linear, one-dimensional spatial structure (Malanson 1980, 1982, Malanson and Kay 1980). This assumption could be made for a variety of riparian settings in the southwestern USA. Several studies provide a basis for such thinking. Irvine and West (1979) cited the isolation imposed by entrenchment of the Escalante River in Utah (although I observed considerable evidence of cattle in their study area in 1976). Turner and Karpiscak (1980) reported differences in vegetation in the Grand Canyon, Arizona. Their study did not consider landscape pattern, but they provided distribution maps for 24 taxa which could be used as a basis for studies of spatial pattern and process.

A third feature of these regions is that, because of low vegetation cover, sediment yield is high and consequently the sediment dynamics of the fluvial systems create conditions which modify the landscape as well as the internal structure of the riparian area. During much of the year the major rivers in these regions carry high loads of suspended sediment. Depositional features are common, but changeable, and sites for colonization may be ephemeral. The longitudinal pattern of particular landforms, with which particular species or plant communities may be associated, can shift along the course of the river, with consequent changes in isolation and connectivity.

Zimmerman's (1969) detailed description of the vegetation of one area of *c.* 1900 km^2 provides a useful illustration of some of the landscape attributes of riparian vegetation in this ecoregion. He differentiated upland species from what were obligate valley-floor species. He noted, however, that in some valley bottoms none of the obligates were found while upland species were present. Four topics of special interest were discussed. First, he noted the differences between perennial and ephemeral streams and among the ephemeral streams. The valley-floor vegetation ranged from closed canopy forests of *Populus fremontii* or *Juglans nigra* along the perennial streams to scattered shrubs or no vegetation at all along some of the ephemeral streams, but among the latter there was a wide range that depended on drainage area above the point of a sample. Second, he noted the longitudinal structure of the vegetation, which included the presence of tributaries as well as distance from the drainage divide. Valley-floor species were best developed in the middle reaches of the channels. Zimmermann (1969) related the longitudinal structure to the relation between ecology and hydrology. He specifically examined the differences that occurred depending on the presence of bedrock or valley fill in the headwater, middle reach, and lower reach zones of the streams. The depth to bedrock controls groundwater depth, which is one of the most important environmental gradients for vegetation in this region. Fourth, he considered the establishment phase as critical to the distribution of valley-bottom species. The requirements for germination and establishment differ in such ways that the combinations of flow regimes and sediment types controlled by the geology and topography differentiate among species and lead to distinct communities within the riparian zone; other factors, such as intolerance of the perennially high water table, were also noted, however. This study, and the riparian range maps of species that are included, provide a stimulating and useful basis for the examination of some of the principles of landscape ecology, and it demonstrates the importance of examining longitudinal structure in developing hypotheses about the ecology of plant communities.

Desert and Mediterranean environments

A considerable number of studies of riparian environments in such regions have appeared in recent years. Walters *et al.* (1980) summarized aspects of species distributions and environment for these environments in the USA. A number of symposia have examined these environments, particularly symposia devoted to riparian habitat in California, which contains both desert and Mediterranean environments (e.g. Sands 1977; Warner and Hendrix 1984; Abell 1989).

Johnson claimed that the major development of riparian ecology has 'happened where riparian ecosystems are most strikingly apparent in the landscape climax pattern – in the North American Southwest' (Johnson and Lowe 1985), and that 'Previously, only a few scattered studies had addressed specific problems in riparian systems' (Johnson and Haight 1984). This argument seems simply to be overstated provincialism. In the area of evapotranspiration and phreatophytes, uniquely significant contributions have been made in the Southwest (e.g. Robinson 1958). This zone has also been notably subject to invasion by exotic tree species originally introduced as ornamentals or for other purposes (*Tamarix*, *Eleagnus*, *Eucalyptus* spp.). Important contributions have been made in plant ecology and in riparian wildlife ecology, but these studies do not preempt those of the rest of the world.

Some of the truly early studies of riparian vegetation were done in the Mediterranean region of Europe and Africa. Tchou (1951) presented an unusually detailed study of the plant association of the riparian zone of southern France. Building on earlier work (citing, for example, Siegrist 1913 and Braun-Blanquet 1915), he related the phytosociological structure of the vegetation, described in extensive relevé tables, to climate, soils, and the level of the water table. He detailed the edaphic factors, especially those related to flooding, and examined the related below-ground structures of several species. With descriptions that seem relatively modern, he referred to mosaic dynamics in response to flooding. He also examined the phytogeographical connections of the species in the riparian zone and contrasted their geographical relations, along with their growth forms, with the surrounding, more xeric,

Mediterranean shrub vegetation. This study could serve as a basis for the extension of current work in the area (e.g. Thebaud and Debussche 1991) to a larger scale.

Structural features of riparian vegetation in desert scrub have seen some attempts at classification for management purposes (e.g. Johnson *et al.* 1984). These studies indicate an awareness of the potential interest in the overall landscape structure, but like those elsewhere do not use the concept to tie together local structures and processes. A notable feature of some studies in this area is the recognition of the longitudinal gradient in the riparian environment where it coincides with a distinct elevation gradient in mountainous areas (Campbell 1970; Pase and Layser 1977). Three studies in New Mexico have examined this gradient. Freeman and Dick-Peddie (1970) compared elevation gradients of east- and west-facing slopes. They reported dominance by shrubs at the lowest and highest elevations observed, with trees dominant in the intermediate elevations. Boles and Dick-Peddie (1983) reported inconsistent pattern species dominance on an elevation gradient. Medina (1986) also reported elevation variation, here related to soil types. Human impacts were considered in all cases. Another area in which the landscape ecology of riparian areas is implicit in the southwestern USA is in considering geographical gradients, the ranges of species, and changes in those ranges. Ford and Van Auken (1982) noted a gradient that coincided with the western range limits of key species in central Texas.

The extreme isolation of incised canyons is not necessary to extend mapping of riparian areas to studies of their landscape ecology. Extensive mapping projects in California provide such a basis. Katibah *et al.* (1984) and Nelson and Nelson (1984) described the efforts of the Central Valley Riparian Mapping Project. Both areal and linear measures of eight major vegetation categories, with two subcategories and three modifiers, were mapped for the depositional bottomlands on 1:24 000 topographic maps. This database could be an excellent resource for landscape studies as well as management. Other data management programs may provide a similar resource. Busch (1984) and Gradek *et al.* (1989) described the Bureau of Land Management system for recording riparian environmental information based on river miles. Stone *et al.* (1984) and Stromberg and Katibah (1984) described efforts to quantify spatial patterns of riparian vegetation at a finer geographic scale, but which can also play a role in landscape studies. In Australia, Bren *et al.* (1988) used a map analysis of internal structure that could also serve as a basis for landscape ecology, and Bridgewater (1980) considered the transverse pattern and its relation to salinity and groundwater; both cases are without the confinement of canyons.

The riparian areas in the southwestern USA have been notable scenes of invasion by exotic tree species (Harris 1966), and the same phenomenon also has been observed in arid Australia (Griffin *et al.* 1989). These invasions are landscape phenomena. Dispersal is the key spatial process; the riparian zone provides the unique spatial pattern. Concepts of landscape ecology run through many of the studies of these invasions, but a true landscape study linking structure and process is yet to be done. I discuss these processes in more detail in Chapter 6.

A few studies specifically examine the landscape ecology of riparian areas in the arid southwestern USA. Harris (1988) described landscape elements connecting geomorphological and vegetational components with the intent to provide a basis for management and studies of landscape dynamics. Szaro (1990) considered the effects of a number of landscape characteristics, including elevation and stream direction, on the distribution of tree species. Szaro and King (1990) commented on the effect that the discontinuous spatial pattern of riparian communities had on the results of sampling methods. These studies indicate that landscape considerations are beginning to be recognized in studies in these regions, but they have not yet plumbed the potential of this landscape.

The question of temporal equilibrium arises in arid and semi-arid regions also. Robichaux (1977) illustrated the changes in the ranges of modern Californian riparian taxa in Miocene, Pliocene, and Pleistocene floras. Williams *et al.* (1984) described another extreme case in which riparian communities are relicts of wetter conditions of the late Pleistocene. Thompson (1961, reproduced in part in 1977) reported the detailed modern history of the riparian vegetation of the Mediterranean region of central California, and noted the early botanical work of Jepson (1893) as well as historical sources to show that human influence has resulted in disequilibrium conditions.

Savanna environments

Hughes (1988) summarized some landscape concepts of work on riparian zones in the savanna and desert areas of Africa. Longitudinal patterns are noted where river systems cross major climatic gradients (e.g. Keay 1949). Elevation gradients also have an effect (Rattray 1961; Lawton 1967; Farrell 1968). Hughes (1988) noted the factors that control the transverse landscape pattern in this environment. While highlighting the importance of depth to the water table, which causes abrupt boundaries of the riparian vegetation, she also considered the difference in this transverse pattern along the longitudinal pattern of rivers where they ran along a gradient of precipitation and flooding. She also summarized the ideas of some authors who documented the role of fire and of a fringe of fire-resistant shrubs in maintaining the transverse landscape structure in these regions (Keay 1949; Hall-Martin 1975; Simpson 1975). Simpson (1975) described both floodplain grassland and a riparian forest 'fringe' which varied from a closed canopy forest to scattered trees in dense shrubs stands. This fire-resistant fringe potentially suffered from overgrazing and might lose its protective qualities. Andrews *et al.* (1975) described the vegetation of the lower Tana River floodplain in Kenya as mostly grass with scattered patches of forest and woodland and areas of shrubland. The mosaic seems to be the result of several forces, including flooding, especially in forest areas, fire, especially in the grasslands, and use by people and animals (e.g. elephants). In Hughes' (1990) own work on the Tana River in Kenya, however, channel dynamics, and consequent geomorphological patterns, contributed to greater overall diversity.

Furness and Breen (1980) produced a detailed map of six plant communities on the floodplain of the Pongolo River in South Africa, which could be used as a basis for studies of landscape processes. Halwagy (1963) produced a similar map while studying succession and fluvial processes on the Nile. Such maps are not unusual, but their linear form is striking. As in desert environments, the rivers in savanna areas sometimes flow in marked topographic incisions. Keay (1949) described such 'steep gullies' in the savanna region of Nigeria. The topographically confining nature of these forests and thickets, along with the physiognomic difference in the vegetation, provide a starting point for the definition of landscape elements. Keay (1949) also noted that in broader floodplains below, and at the very headwaters above these gullies, the vegetation changed to a broader and more open *Acacia* savanna. His diagram (Keay 1949, p. 355) of this landscape pattern reveals a recognition of landscape concepts.

Wyant and Ellis (1990) examined floristic patterns at both a regional and a landscape scale in northern Kenya. For an analysis of regional pattern they considered riparian vegetation on both east and west sides of a mountain range. They used this geographical distinction along with north–south trends to produce a diagram of species richness in four quadrants corresponding to gradients in precipitation and fluvial activity. They found differences in composition based on the scores for stands in detrended correspondence analysis and differences in species–area curves. For the landscape scale they analyzed the differences in species composition with elevation along the channels. Regressions of both species richness and DCA scores on elevation were significant in only two cases, those on the west, and richer, side of the mountains. They concluded that regional trends and local site characteristics, such as substrate and fluvial dynamics, were more important than the longitudinal effect of elevation along channels.

The contrast between this elementary landscape approach in Africa with the aspatial functional approach in America is noteworthy. Some reviewers might attribute this to a greater concern with the natural history and biogeography of an areas that has seen relatively little study compared with the USA, where numerous practicing ecologists must vie for new approaches, but it should reaffirm the idea in landscape ecology that natural history is important.

In summary, as Johnson and Lowe (1985) noted, in arid and semiarid environments, riparian areas are visually distinct, but this visual distinction also means that principles of landscape ecology can be tested here. In general, the transverse pattern is simplified, so that the longitudinal pattern is emphasized. Sinuosity or curvilinearity is clear: where the river (and confining canyon) bend, the narrow band of riparian vegetation also bends, unlike in some areas where the river meanders across a wide and straight floodplain. Connectivity is also clarified. In some areas riparian vegetation grows only on point bars, and the river is against a sheer canyon wall in other places (Figure 3.4). The fact that some streams are intermittent and/or ephemeral also presents opportunities for the analysis of connectivity. In these dry areas the dimensions and shapes of riparian elements are relatively easy to discern, and so can be quantified. Riparian areas in arid ecoregions should be targeted for studies of the principles of landscape ecology.

Arid region riparian environments provide unique opportunities to develop and test hypotheses in landscape ecology because of the distinctness of their spatial pattern. Concepts related to linearity, such as sinuosity and width, can effectively be developed in such places.

Tropical forests (Rainforest Division)

The most notable feature of the landscape structure of riparian environments in the tropics is their extent. In South America, in Africa, and in southeast Asia and the nearby islands, major rivers traverse areas of

Figure 3.4 Within incised canyons, the vegetation of one point bar can be spatially isolated from that of another by the topography; Harris Wash, Utah.

tropical rainforest. Where the rivers meet the ocean, extensive deltas have developed in some areas; here some of the tidal influence on the environment includes salt water, and the vegetation can be called mangrove, and is thus beyond the scope of this book. Throughout the river basins, lowlands can be flooded by precipitation as well as by river flooding, and thus the boundary between riparian and non-riparian swamp is difficult to determine. None the less, some major features of riparian landscape ecology can be considered. Landscape ecology has been a specific paradigm used to approach tropical riparian environments, but only in the sense of the regional interaction of landforms and ecological processes. Other aspects of research in these areas reveals that landscape concepts are implicit and that a more open landscape approach including specifically spatial relations may be worthwhile.

Although directed toward a different purpose, the description and data on floodplains provided by Welcomme (1979) is a useful source (some of this information applies to savanna as well as rainforest systems). He compiled information on the extent of many floodplains in the tropics. He listed total areas inundated at peak flood (e.g. 80 000–100 000 km^2 of the Gran Pantanal of the Paraguay River and 92 000 km^2 of the Sudd of the Nile River), and presented maps of major floodplain locations in Africa, Asia, and South America. He defined three major types of floodplains: fringing, which are the linear riparian corridors; internal deltas, where topographic plains may flood from a combination of precipitation and over-bank flow from a river; and coastal deltaic floodplains. In South America he mapped extensive floodplains along the Parana and Amazon Rivers. In Africa, many rivers contain both delta and fringing floodplains; notable fringing floodplains are on the Zambezi, Senegal, Niger and Volta Rivers. He noted that Zaire River contains the only extensive forested floodplains similar to those found in the Amazon basin, but found that the area was difficult to assess, in part because of very low population density. Although the Asian rivers are well known for their extensive deltas (e.g. the Ganges–Brahmaputra delta, forming the sundabarans, and the Mekong), little work has been done along their lengths.

Much of the work in tropical areas has been basically floristic. The identification of species and the identification of communities on floodplains has been examined in some detail (e.g. Balslev *et al.* 1987). Recent reviews have noted this starting point (e.g. Kahn and Mejia 1990; Myers 1990; Paijmans 1990). Work has focused on the relation between flooding conditions and the assemblage of species found (e.g. Campbell *et al.* 1992). Klinge *et al.* (1990) provided an excellent summary of these efforts in tropical South America. This and other studies (Prance 1979; Klinge and Furch 1991) emphasized the hydroperiod and water chemistry of flooding as the key factors to discriminate among forest types. Kubitzki (1989) discussed regional differences in the floristics of Amazonian inundation forests and related these to their evolutionary history. Work on African riparian rainforests is scanty, but Colyn *et al.* (1991) have noted some aspects of rainforest biogeography related to the river as a barrier (see Chapter 6).

Three features of tropical riparian landscapes are notable. First, the location of floodplain sites within the basin determines in part the differing hydroperiods and water chemistry of flood flows. Second, the development of floodplains sites and the diversity, assemblages, and species dynamics are related to the geomorphological dynamics of these major alluvial rivers. Third, the connectivity of floodplain sites, especially through specialized means of seed dispersal, is a notable feature of the landscape structure of tropical riparian habitats.

Flood regimes and water chemistry

Flooding in the tropics in South America has been identified as seasonal, primarily on large floodplains, or irregular, on the small floodplains of low-order rivers. The vastly more extensive areas of seasonal flooding are further subdivided by the annual pattern of flooding, which may be unimodal, bimodal, or polymodal. The regime of a given region is determined by its latitude and thus its relation to the Inter-Tropical Convergence Zone and seasonal precipitation, and also by its position in the basin and the degree to which pulses of precipitation are modified or increased by the position of the site relative to riverine influences.

As with most large rivers, the hydrograph of the Amazon is flatter and more expanded in lower reaches of the river. Goulding (1980) noted that for tributaries, rainfall maxima, coinciding with the high sun period on opposite sides of the equator, are months apart. These distinct flood peaks are combined in the main trunk of the Amazon where waters from both directions combine. One factor noted by Goulding (1980) is that high water in the main trunk can cause backwater flooding in the tributaries. The hydroperiod thus has a geographic distribution which may affect the spatial relations of riparian rainforests. Welcomme (1979) showed hydrographs of 14 major rivers, and illustrated the longitudinal variation in the hydrograph for the Niger, Senegal and Chary Rivers in Africa.

Goulding (1980) has given a readable explanation of the hydrology of the Amazon rivers. Those rivers that rise in the Andes carry large sediment loads, appear turbid, and are called 'whitewater'. Those rivers that rise in the Brazilian and Guiana Shield areas carry low sediment loads and are nutrient-poor; these rivers are called 'clearwater'. Rivers rising in Tertiary lowlands are nutrient-poor, very low in sediment, but contain dissolved acidic compounds and are called 'blackwater'. Because clearwater rivers also arise in some of the same conditions as blackwater rivers, the causal factors are not clear.

Types of 'inundation forests' have been variously defined in Amazonia. One distinction is based on water chemistry: varzea is forest flooded by whitewater rivers; igapo is forest flooded by clearwater and blackwater rivers. Another distinction is based on hydrology. Prance (1979) delineated seven types of inundation forest, as follows:

Permanently flooded:
1. white-water swamp forest
2. permanent igapo (gallery forest)

Periodically flooded:

3. floodplain forest
4. seasonal igapo
5. mangrove

6. seasonal varzea (with or without associated grassland)
7. tidal varzea

Irmler (1977) distinguished between varzea and igapo as inundation forests in which the period of flooding was very regular, and floodplain forests found at the higher reaches of rivers where flooding was an irregular response to more local precipitation events. He used stream benthic fauna to distinguish water types and thus three types of varzea and two of igapo. Klinge and Furch (1991) related water chemistry to biomass chemistry in order to distinguish varzea and igapo.

Because of the relations between water chemistry and the source areas for rivers noted by Goulding (1980), the chemical distinctions affect the spatial distribution of igapo and varzea. This aspect of landscape pattern has not been investigated. Various rivers have been identified as to their 'color', but I have not been able to locate any map of the distribution of these two forest types.

Geomorphological dynamics

Different areas in the Amazon basin contain different features of fluvial geomorphology. Because the fall of the main trunk of the Amazon over 2000 km is only 80 m, the river contains many meanders (even though the river has not fully adjusted to Pleistocene alluviation, see Sternberg (1975) cited by Goulding (1980), and so might be analogous to paraglacial conditions in mountain environments) and braids. The main tributaries also have forms typifying large alluvial rivers. Campbell and Frailey (1984), Salo *et al.* (1986), Frost and Miller (1987), and Rasanen *et al.* (1987) have considered that the geological and related geomorphological processes of river dynamics have contributed to the development of species diversity in the Amazon basin, but these and other studies address internal structure and will be reviewed in Chapter 4.

Regardless of the relative importance of fluvial dynamics in the biological diversity of the tropics, it is likely that the dynamics do play an interesting role in the spatial configuration of the landscape elements produced. The relative position of accreting sites in close proximity to each other, and the density and sizes of such sites within larger regions, may play a part in determining the degree to which they expand the niche space for pioneer species and the extent of differentiation that could occur and lead to speciation. Sioli (1984) and Salo (1990) have discussed the relation between geomorphological dynamics and landscape pattern with specific reference to landscape ecology. They considered the importance of the interaction without reference to specific patterns of configurations. While demonstrating that important patterns which affect ecological interactions originate in geomorphological processes, these authors do not apply a truly spatial perspective.

Connectivity

The landscape considerations of proximity, density, and position extend not only to early successional stages, but also to the riparian forests in general. The dispersal of species, and thus the genetic diversity within species in the tropics, is a question of general biological interest. The individuals of tropical rainforest species are relatively isolated from their relatives. The dispersal of propagules from one forest or region to another is interesting in all riparian environments, but in the Amazon Goulding (1980) has examined an interesting relation between fish that live in the riparian forest when flooded, eat seeds, and disperse them to other areas (see Chapter 6 for an expanded discussion). Proximity, density, and position of the trees and forests in question, along with the spatial behavior of the fish, will need to be examined in order to fully develop the study of this relation.

It will be difficult to test the principles of landscape ecology in riparian rainforests because of their floristic complexity and their great extent, which mean that boundaries, and thus shapes, areas, and connectivity, will be difficult to determine. Nevertheless, because of the importance of these areas for global diversity and biogeochemical cycling, and the usefulness of landscape ecology in management, these problems should be tackled. The spatial distinction between varzea and igapo seems a good starting point. The mosaic as a concept, as opposed to linear dimensions, could effectively be considered in this environment.

Subtropical floodplain forests (Humid Temperate Domain, Subtropical Division)

Although only one type in Bailey's (1989) formulation (with analogous ecoregions in southern Brazil, China, Australia, and New Zealand), this forest type contains two of the ecoregions that he defined earlier for the USA, a southern mixed forest and a specific southern floodplain forest (Bailey 1976, 1980). This latter type is riparian forest, and has seen extensive study. These riparian forests of the Mississippi embayment and some other southeastern USA sites are distinct because of their lateral extent. While all riparian zones tend to be linear, these are relatively wide. Exceptions occur northeastward along the Atlantic coast, where

Glascock and Ware (1979) described riparian forests as sharply defined by steep slopes and distinct vegetation. In the Mississippi embayment the extensive alluviation has resulted in larger areas of frequent flooding and saturated conditions of long duration. The warm climate, plentiful rainfall, and low relief make these forests less visually distinct from surrounding uplands than cool deciduous forests, but even more ecologically distinct. Here are found the species most tolerant to flooding (*Nyssa aquatica* and *Taxodium distichum*) (Teskey and Hinckley 1977*b*). Because of the spatial extent of the southern floodplain forest, the geographical range of some species can be seen directly in relation to it. Little (1971, 1977) mapped the range of 94 conifers and 106 important hardwoods in the USA, and 166 minor hardwoods in the eastern USA. The Mississippi embayment is a notable feature on several of his maps (e.g. Figure 3.5). I visually assessed these maps, and classified species ranges relative to the Mississippi embayment. I classified both positive and negative spatial association, and whether the distinction was strongly or weakly related because of widespread distribution elsewhere, patchy distribution, or a range boundary on only one side. One important hardwood, *Nyssa aquatica*, has a strong positive association; one conifer (*Taxodium distichum*, widespread in the southeastern USA), two important hardwoods (*Carya aquatica* and *Catalpa speciosa*) and four minor eastern hardwoods (e.g. *Gleditsia aquatica*) have weak positive associations here. Four conifers (e.g. *Pinus taeda*), 19 important hardwoods (e.g. *Carya cordiformis*) and 11 minor hardwoods (e.g. *Aralia spinosa*) have strong negative associations, and one conifer (*Pinus glabra*), six important hardwoods (e.g. *Liriodendron tulipifera*), and 21 minor hardwoods (e.g. *Prunus caroliniana*) have weak negative associations. This distinctive biogeographical pattern indicates that the boundary of the embayment is a distinct ecotone at a coarse scale. This aspect of the transverse landscape structure should be investigated.

Langdon *et al.* (1981) defined eight physiographic classes, based on geomorphology and hydrology, in which bottomland hardwood–cypress forests exist in the southeastern states. They noted the productivity and commercial values of the lands and suggested research and management linkages. Turner *et al.* (1981) presented a map of 12 types of wetland in the central and southeastern USA, five of which might be considered as riparian forests. Their map does not include the riparian zones of the upper Mississippi and Ohio drainages, however, because of the small scale. The 'Southern floodplain forest' and 'Pocosin' are well represented. The authors highlighted the impact of land use changes on

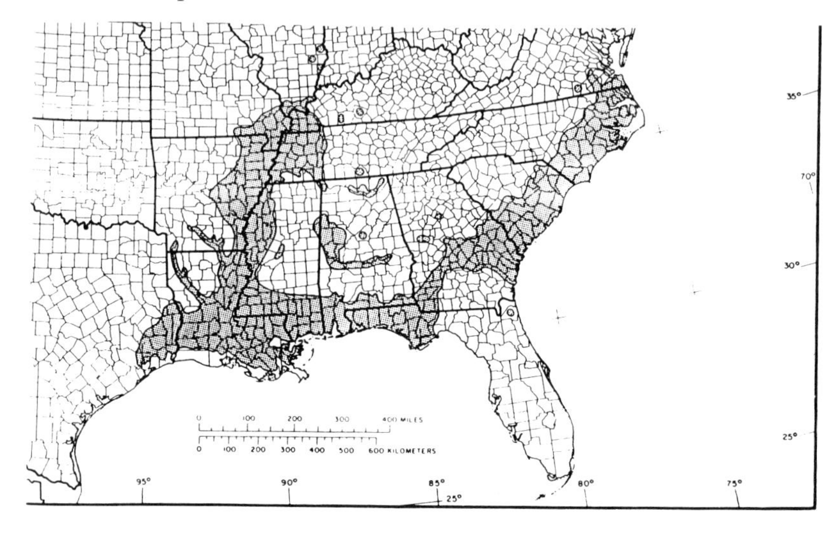

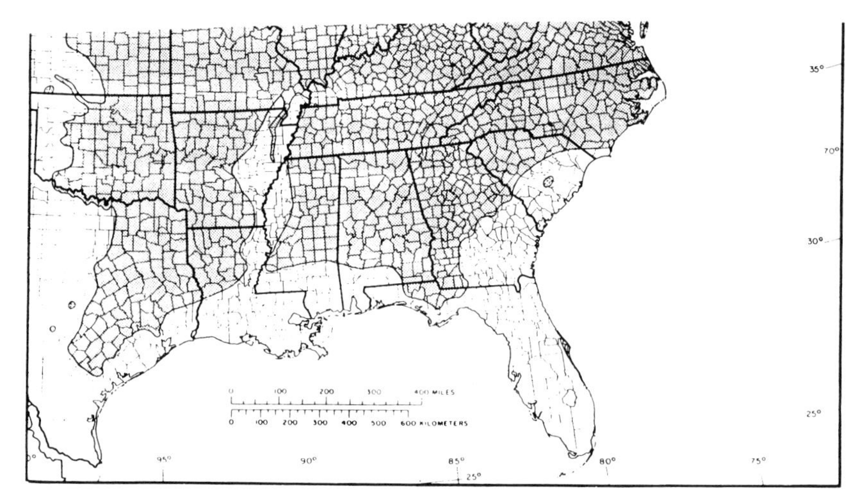

Figure 3.5 Maps produced by Little (1971) show the distinct phytogeographical expression of the Mississippi embayment.

the areal extent and landscape structure of these forest types, especially the former. One can infer that the process of conversion has been variable and that the resulting landscape is one where the very broad corridor of the region has been reduced to narrow corridors and patches. Where it remains broad, the riparian forest corridor is inversely opened up by patches of development and road corridors. In Florida, extensive areas of floodplain forest exist because of the generally low topographic relief. These floodplains are dominated by two forest types, a *Quercus*-dominated hardwood type and *Nyssa aquatica – Taxodium ascendens* swamp. Large-scale mapping projects here (e.g. Pratt *et al.* 1989) are beginning to provide a basis for landscape-level studies.

These bottomland hardwood forests have been noted for their biomass, attaining over 75 m^2 ha^{-1} of basal area in several cases (Hall and Penfound 1939, 1943; Good and Whipple 1982; Duever *et al.* 1984; all compared by Brinson 1990) and over 600 t ha^{-1} in one case (Duever *et al.* 1984). Brown (1981) reported very high levels of primary productivity and respiration in this environment. Many of the detailed studies in this region have been on the effects of flooding on trees or the engineering hydrology of the fluvial system. These topics are examined in subsequent chapters.

The landscape structure of the extensive forests of this region presents both opportunities and problems. At most scales at which analyses have been done, the landscape disappears because the processes of interest operate within a single element, perhaps within a single tessera. Because the riparian element is not so distinct, its interactions with surrounding elements are difficult to discern and it may be difficult to design studies that directly reflect on the principles of landscape ecology. The extensive deforestation in this area, however, may provide cases where the landscape pattern becomes apparent.

In another related area, the wetlands of the Mississippi delta have received study that considers both their own spatial pattern and their interaction with other environments. In particular, their dependence on the balance of a sediment budget between inputs from the Mississippi river system and encroachment by the sea has been observed. Baumann *et al.* (1984) noted that the balance depended on both the particular nature of the hydrological relation of wetland and river and the rate of submergence by the sea. Neill and Deegan (1986) and Rejmanek *et al.* (1987) examined potential habitat change and successional trends associated with delta sedimentation and erosion. Kesel (1988, 1989) noted that the sediment load of the Mississippi river had declined over the past century owing to changes in land use and the construction of dams on the major tributaries. The result has been a major loss of delta wetland area. This far-reaching connection illustrates the scale at which landscape ecology could be applied to the riparian environment of the entire region. As in the rainforest region, the mosaic concept would be a starting point, but not to the exclusion of linear concepts.

Humid broadleaf forests (Humid Temperate Domain, Warm and Hot Continental Divisions and Marine Division)

The transition from the subtropical ecoregion to the humid continental forest has been noted in terms of the floristic boundaries. Robertson *et al.* (1978) described the riparian forests in the area called the northward limit of the southern forest. They did not, however, report any particular biogeographical consequences of the location, but instead focused on the internal structure. Lindsey *et al.* (1961) used phytosociological analyses in the floodplain forest to show a continuum along a 370 km latitudinal gradient, identifying regional climate as a controlling factor. This transition follows the Ohio River northeastward. In this specific region, Sampson (1930) included floodplain sites in his discussion of succession in swamp forests in Ohio, and Lee (1945) quantified the vegetative cover on floodplain sites in Indiana. On a limited geographical scale, Ericsson and Schimpf (1986) reported changes along a 16 km stretch of river in northern Minnesota, which they related to river gradient. The same ecoregion province exists, however, in Poland and Czechoslovakia, and comparable forest types can be found on the eastern coast of the USA, in western Europe, and in East Asia (Bailey 1989). In general the mixed conifer–hardwood forests of northeastern North America present a continuous canopy over large areas. The riparian zone does not markedly alter this visual landscape. Nevertheless, these zones are ecologically distinct. A host of studies indicate their intrinsic interest. Most of these studies are oriented toward either internal structure, particular processes, or both, and will be discussed in subsequent chapters. It is worth noting here, however, that a considerable portion of pioneering ecology has been carried out in these riparian forests, both in terms of concepts (Buell and Wistendahl 1955) and methods (Sigafoos 1964).

In Europe, floristic descriptions for many regions are available from a single symposium (Gehu 1980), but landscape structure was not a general topic. A large-scale study in this riparian forest type in Czechoslovakia, sponsored by the International Biological Programme and Man and the Biosphere, has also been summarized. Ecosystem processes, such as primary productivity, were the major focus of this effort (Vasicek 1985), but many details of processes give information relevant to landscape ecology. Notable points include the description of microclimate, soils, and exchange processes before and after river control structures. These studies carefully characterized the floodplain forest ecosystem, but, as is typical of ecosystem studies, did not consider its landscape structure within the surrounding matrix (e.g. Uhrecky *et al.* 1985; Vasicek 1985; Palat 1991). Because the major change in hydrography is on a landscape scale, however, these studies may serve as a basis for future landscape ecology.

Another major team approach in France has emphasized aspects of the interaction of fluvial geomorphology and floodplain ecology (Amoros *et al.* 1987*a*). Bravard (1987) incorporated a landscape approach in his examination of the Rhone between Geneva and Lyon. While much of his attention was devoted to specific processes at specific places, his attention to the entire section of the river embodies landscape concepts. He identified three geomorphological units which control the longitudinal structure, including the fluvial dynamics within four longitudinal sections. This approach, accompanied by numerous maps, diagrams and photographs, shows that the longitudinal structure affects the flows along the corridor and the transverse structure, which is affected by those flows. Particular landscape features of note are the complex channel patterns and abandoned channels at La Chautagne (just below the gorge at Bellegarde), Basses Terres (site of a major channel change), and below the confluence with l'Ain (on extensive proglacial gravel deposits). Bravard (1987) also notes the landscape effects of human modifications of the river system, including the recognition of the effects of hydraulic structures in one place affecting other places being embodied in law. Part of the transverse pattern is also determined by the geomorphological setting. Pautou *et al.* (1979, 1985) related vegetational patterns along the Rhone to the spatial relations of braided and meandering reaches as well as hydroelectric plants, and Roux *et al.* (1989) summarized the historical changes in the ecology of the system since 1750. Rivers at other scales have also been studied: on a major tributary of the Rhone, the Ain, Bravard (1986) and Roux (1986) again noted the relation between the fluvial processes and the spatial configuration of the ecological units, and Girel and Manneville (1991) have described the vegetation of low-order streams in the region. Pautou and Girel (1986, 1988) have adopted a specifically spatial and landscape approach. Other aspects of studies on the Rhone and Ain are discussed in following chapters.

The French team approach has become specifically focused on landscape ecology. Henri Decamps (cf. Decamps 1984; Decamps *et al.* 1988) was elected president of the International Association for Landscape Ecology in July, 1991. In his plenary address to the IALE congress, he cited may of the themes of this book, most notably that riparian landscapes are excellent areas for the future study of landscape ecology because of their linkage between spatial pattern and process. Tabacchi and Planty-Tabacchi (1990) and Tabacchi *et al.* (1990) conducted an explicitly longitudinally oriented landscape study of riparian vegetation. They examined the longitudinal pattern in species richness and floristic continuity along the Adour River, southwestern France. Their method of examining floristic affinity by a moving window of consecutive sites provided insights into the turnover of species and distinguished distinct longitudinal zonations related to climatic and disturbance factors and tributary effects, and they noted that the upstream and downstream patterns in continuity are not symmetrical. The concept of connectivity between the Pyrenees source and Atlantic end of the river corridor was discussed.

The French teams working at Toulouse and at Grenoble and Lyon both have an excellent resource for landscape ecology in detailed maps of the riparian areas in question. The group at Grenoble has a particularly diverse set of maps at different scales. Notable among these map products are the work of Pautou *et al.* (1979) at 1:25000 for sections of the Rhone and of Girel *et al.* (1986) for the Ain, a tributary of the Rhone, which, though printed at 1:10000, was compiled at 1:5000. This later map served as a basis for several investigations of riparian ecology (see Roux 1986). Mapping of the upper Rhone valley at 1:5000 is now in progress (J. Girel, personal communication). If studies in landscape ecology are to progress, they must be based on an accurate mapping of vegetation and other ecological features. These French teams now have the best basis for posing and testing hypotheses in riparian landscape ecology, and from this spatial base interesting studies should be forthcoming. Ongoing research related to the landscape structure includes the study of the historical redistribution of soil and water through distributaries along the Isere, which has led to a change from bands of vegetation type parallel to the river to intermingled patches (J. Girel, personal communication), and the study of the c. 50 yr post-flood recovery along the course of the Tech, wherein longitudinal effects are apparent (M. Izard, personal communication).

Elsewhere in Europe, phytosociological classifications have been employed in studies of this riparian type. For example, Hermy (1980) classified riparian woods in Belgium, and Curry and Slater (1986) improved on this aspect in a classification of riparian vegetation in Wales. They used floristic data to classify five major and nineteen minor types of riparian vegetation, which they related to environmental variables. Curry and Slater (1986) also classified riparian vegetation and found evidence of longitudinal structure, in their case related to elevation, among other variables such as slope and substrate. Gehu and Franck (1980) also reported a gradient in *Salix* forest structure with elevation in northern France. From another viewpoint, Wildi (1989) reported that a continuum in vegetation structure was disintegrating, due to human influences, into floristically distinct groups. This analysis implies that changes in both the longitudinal and internal structures are not static. Mason and Macdonald (1990) reported vegetation variation throughout several basins, but addressed management practices and not longitudinal patterns. Slater *et al.* (1987) showed how such information can be applied in a conservation program. Although their mapping of vegetation types is useful, it did not address the special spatial constraints of these corridors. Carbiener and Schnitzler (1990) described the longitudinal gradient in vegetation structure and function along a 300 km smooth profile of the Rhine. They reported a downstream gradient in floristics, spatial pattern of tesserae, diversity, biochemical cycling efficiency, and the breadth of ecological tolerances or niche widths. Although not phrased in the vocabulary of landscape ecology, this study provides the most comprehensive analysis of multiple changes along the longitudinal profile of a riparian environment.

Hupp (1982) described longitudinal variation along a short section of a small stream in a hilly area of Virginia. He noted differences between species growing where a broader floodplain could develop and those where the stream cut a more incised channel through resistant bedrock. Hupp (1986) examined low-order streams in the same area and noted that as one approached the headwaters of streams fluvial landforms and riparian indicator species disappeared.

The riparian environments of this ecoregion type have been studied extensively and intensively. These studies provide an excellent basis for the new work incorporating a landscape perspective. Both linear and mosaic approaches can be successfully applied in this environment. Current developments in this area should prove richly rewarding in the next few years.

Forest-grassland transition and grasslands (Humid Temperate Domain, Prairie Division and Dry Domain, Temperate Steppe Division)

In the USA, north and west of the Mississippi embayment, the rivers flow from drier regions. The smaller drainages such as the Iowa and

Illinois rivers reach into prairie but are still bordered by a riparian forest; the major Arkansas and Missouri rivers flow from subtropical steppe mountains (see below) across subtropical steppes and prairies (Bailey 1989). In some areas of the agricultural grasslands, where wet soils or flooding have precluded the plow, the riparian forest is only a remnant of the former forest. Still, much of the riparian forest land has been drained and plowed (see maps in Turner *et al.* 1981, p. 26). What is left has been grazed and selectively harvested for timber. Here the riparian forests visually appear as corridors. Maps from the days of early agricultural settlement in the nineteenth century reveal that woodlands were almost exclusively riparian in nature in much of this region (Figure 2.7). As noted above, where broken up by farmland they resemble the chain of an archipelago in a sea of corn, wheat, or beans (Figure 2.6). Attention focused on the riparian vegetation of this region in the USA from an early date (Ridgway 1872; Macbride 1896; Fitzpatrick & Fitzpatrick 1902; Gleason 1909; Aikman 1930; Sampson 1930; Turner 1936; Howard & Penfound 1942; Lee 1945).

Another early report on riparian vegetation is from this ecoregion type in the Soviet Union. Illichevsky (1933) reported that species were segregated according to differences in period of inundation and soil type related to flooding. While he did not consider the longitudinal structure of the rivers, except to note that often there was no change from source to mouth within what he called a belt, he did present an early conceptualization of the relations of vegetation with geomorphological features. In this type of environment the riparian forest vegetation grades from being ecologically and floristically distinct from upland vegetation to conditions where it is the only woody vegetation and is therefore visually distinct as a corridor higher than its surroundings. At this point, some features are shared with those of semi-arid regions, and in fact the climate too grades into the semi-arid. Gareera and Naumova (1980) specifically concentrated on the longitudinal change with elevation above sea-level of riparian vegetation in a steppe region in the Mongolian People's Republic. The purpose of their floristic analysis was to define territorial units along the longitudinal profile of rivers.

The original condition of these riparian areas in the prairie of the USA Great Plains region is disputed. Gleason (1922), working with much less data than now available, discussed the development of midwestern USA vegetation in the Holocene. He noted the importance of the landscape structure of the river valleys for the diffusion of tree species. He specified a hydrarch series of plants that rapidly advanced along stream courses into what he thought of as a prairie dominated area, and that the

> forests therefore soon assumed the form of long branching strips, following the rivers from their mouths toward their sources. Through the slow lateral advance these strips were widened and eventually became confluent across the interfluvial prairies, frequently isolating portions of prairie in the process. . . On the other hand, the wide alluvial bottom lands of the larger rivers, notably the Missouri, Mississippi, and Illinois, seem to have resisted forest invasion, and on them the forests were limited to relatively narrow strips along the channel and the abandoned oxbows, alternating with strips of prairie. This condition, recorded by numerous early observers, is so unlike our modern experience with floodplain vegetation in the Middle West that it at once stimulates inquiry into the circumstances of more recent forest development.

Gleason (1922) then went on to make an argument from observations of the spatial pattern of vegetation and topography that fire, from which vegetation on east banks of rivers is partially protected, resulted in the pattern found. Weaver *et al.* (1925) also concluded that the landscape structure of riparian vegetation in Nebraska was the result of processes of species movement across the landscape.

For the western part of this region in the USA, a common assumption is that the primeval vegetation was comprised of gallery forests made up of the same species now found as dominants (i.e. *Salix* spp. and *Populus deltoides*) (Teskey and Hinckley 1978*c*). Historical analyses have reported that considerable areas had no such gallery forest, however (Crouch 1979*a*,*b*; Lindauer 1983; Knopf and Scott 1990). Knopf (1986) wrote that the east–west rivers across the Great Plains would have had vegetation during glacial advances, which then presumably disappeared during the interglacial stages. He hypothesized that these intermittent corridors affected the evolution and biogeography of bird species in the region.

Further east in the USA, in the transition zone, the riparian floodplains probably were forested, but here the question of the processes creating the riparian pattern arises. Baker *et al.* (1987, 1990) reported the development of riparian vegetation in eastern Iowa over 2500 yr. Their palynological analysis indicates a surrounding environment of floodplain and valley margin forests with some open environments in the region during the period from 2500 yr BP until historic times. Pollen and macrofossils indicate that the abandoned oxbow from which they sampled changed from wet fringe plants such as *Zizania aquatica* toward a marsh dominated by sedges. Chumbley *et al.* (1990) also examined riparian sites in eastern Iowa, but for 9000 yr. They reported a change from a spruce parkland environment to a deciduous forest type, which may have been similar to present communities, between 9000 and 6000 yr BP. In the period around 5000 to 3000 yr BP the tree species were largely replaced by prairie forbs and grasses. A mesic oak forest then returned and dominated up until historical times. In this case at least, it seems that riparian forests have been present with many of the same genera for several millennia, but that they were not immune from displacement during very dry periods. It is noteworthy that during the dry period these areas did not see an influx of *Salix* spp. and *Populus deltoides*, which now dominate the riparian zones of the drier prairie region to the west.

The importance of fire in creating the landscape structure in this region has been discussed in some detail. In the eastern regions of this area, it is clear that trees can establish and grow on the uplands. In addition to the success of planted woodlots, abandoned fields now succeed to forest and efforts to maintain or recreate native prairie find that the establishment of trees must be combatted, usually by fire. It is usually assumed that the riparian zones avoided burning because their fuel loads were not conducive to continuing combustion and they are topographically protected where fire would be less likely to spread down river bluffs and onto the floodplain. The vegetation is not dry, the understory plants are not very flammable (present day understory plants are mostly *Rhus radicans*, poison ivy, and *Urtica dioica*, stinging nettle, but some flammable fuels such as *Galium aparine* do accumulate by late summer). The topographic effect is based on observations of fire behavior in other regions. In the western regions this same argument has been made for some areas, but some riparian zones are not as topographically distinct. The Platte River system is in a broad valley, but the riparian forest is maintained only along the river; it may, however, be relatively recent (Crouch 1979*a*,*b*).

In the past the transverse pattern of the riparian zone has been the primary component of landscape structure that has been considered in these regions. In this visually distinct environment the landscape linkage of riparian sites (and their internal structure) with the hydrological and sedimentological regime of the river is clearly related to the principles of landscape ecology. These principles have not, however, been worked out in relation to the spatial configuration of these riparian elements in relation to those regimes. In these particular cases, the location of the dams and the geomorphological and hydrological changes that have resulted are relatively straightforward and could be analyzed spatially in relation to the vegetation. While the longitudinal pattern in the USA has not been discussed, the transverse pattern has been mentioned in terms of the gradient from forested land into grassland (Aikman and Gilly 1948). Weaver (1960) described the landscape pattern of floodplain vegetation of the Missouri River, primarily in Iowa, Nebraska, and South Dakota. He noted the topographic control of the transverse pattern and, in terms of tributary streams, the longitudinal pattern. He also noted the floristic effects found because several riparian tree species reach their western range limit in this area. Gesink *et al.* (1970) documented both the longitudinal and transverse structure of riparian vegetation along the Arkansas River in Kansas by running line transects perpendicular to the channel along a 400 km reach. While dominant and associated tree species changed over this length, the more detailed transverse structure was related to channel dynamics through their influence on internal structure, as discussed in Chapter 4. Rex and Malanson (1990) examined the shapes of remnant patches in this area. We found that a fractal shape index was most strongly associated with the degree of human impact. More extensive studies of the remaining riparian forest could be done with more attention to the transverse structure and patch characteristics as they change longitudinally along the course of the river.

As in more arid regions, the emphasis on the linearities of the visibly distinct riparian corridor would seem to be a fruitful approach for landscape ecology in this region. Mosaic processes will also be found to be important, however, because the riparian zone is set in a more complex landscape including varied land uses.

Mountains (Dry Domain, Temperate Steppe Regime Mountains and Humid Temperate Domain, Warm Continental Regime Mountains and Marine Regime Mountains)

The major western tributaries of the Mississippi, the Missouri and the Arkansas, rise in the Rocky Mountains, as do the Colorado, flowing into desert, and the Columbia, flowing into the west coast oceanic forests. Studies in the coniferous forest tend to have one of two foci: either on the vegetation of gravel bars or on the functional role of the riparian zone in acting as a buffer between aquatic and terrestrial systems. This latter focus reflects a concern with logging operations. The former focus has been taken up on the larger rivers of the northwestern USA, Alaska, and British Columbia where major hydrological events and processes (e.g.

Figure 3.6 (*a*) In mountainous areas streams can cut narrow valleys, Kings River, California, or (*b*) flow through wide U-shaped valleys carved by Pleistocene glaciers.

floods) create landscape pattern (e.g. Fyles & Bell 1986). The latter focus has been on small streams and the effects of specific structures such as log steps (e.g. Marston 1982). Again these studies tend to focus on particular processes or ecosystem functions and only in passing refer to the landscape role of the riparian zone.

These riparian forests are often very narrow along downcutting streams in mountain valleys (Figure 3.6*a*). The ecological distinction may appear only among the understory species, as the dominant trees may also be those generally found on mesic sites and not specifically on riparian sites. In other areas, however, either in the U-shaped valleys of

former glaciers or on glaciofluvial deposits, a wider, flatter valley is available, and more specifically riparian vegetation grows (Figure 3.6*b*). In these areas a strip of deciduous shrubs and small trees, often dominated by *Salix* species, appears within the coniferous forest (e.g. Marston and Anderson 1991). The spatial character of these riparian areas is controlled by the geomorphology of the area, but this may be extensively modified by the work of beaver (*Castor canadensis*). The two types of riparian areas may alternate along the same stream. Resistant bedrock may lead to a constriction where downcutting predominates, while upstream and downstream the river meanders or forms pater-noster lakes. The distinct spatial characteristics determined by these features may be useful for the study of principles of landscape ecology because the quantification of spatial structure will be relatively straightforward. More than two types may exist: Evenden (1989) described 28 riparian plant communities along four streams in the Trout Creek Mountains, Oregon.

Baker (1989) examined riparian vegetation in the Rocky Mountains of Colorado in terms of macro-scale controls such as the pattern of glacial outwash and micro-scale influences such as local topography and flooding. These differences in geographical scale indicate the potential complex landscape structure in mountainous environments. The longitudinal structure in this environment can be complex because of control

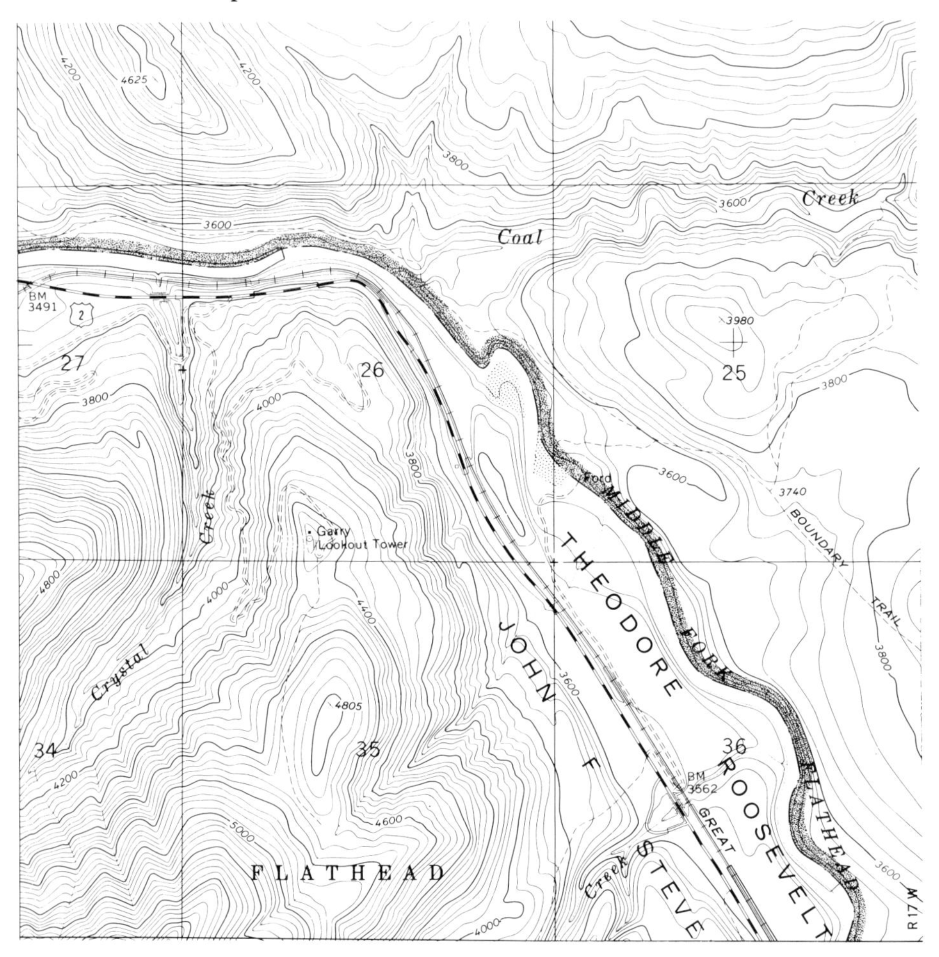

Figure 3.7 Map from USGS, Stanton Lake, Montana, illustrates that narrow and broad valleys can alternate in mountainous areas depending on the geological structure.

by bedrock structures. A river may be confined to a narrow canyon, widen into a relatively broad floodplain with active alluvial deposits, and return to a confined canyon within the space of a few kilometers (Figure 3.7). The transverse pattern is obviously controlled by such changes in the geological foundation. Fonda (1974) reported the effects of river terraces on vegetation patterns in Washington, and Kovalchik and Chitwood (1990) illustrated different transverse patterns that result from different combinations of landforms, soils, and vegetation in riparian areas in the mountains of Oregon. The temporal aspect of these riparian areas is noteworthy. Most of these areas are paraglacial (*sensu* Church and Ryder 1972), i.e. still affected by glacial processes in the Pleistocene. These effects are seen in massive amounts of sediment created by glacial erosion that are still filling valley bottoms (Figure 2.11). These deposits, often coarse, become active during large floods and are temporarily stored in gravel bars, which form the basis for much of the riparian ecology. In mountainous regions of the west coast of the USA, which Bailey (1989) classifies separately as marine regime mountains (with similar ecoregions in western Europe, the Andes, and New Zealand), interactive processes of geomorphology and ecology are also noted. In particular, these mountains have been the scene of major work on woody debris. Much of this work is focused on internal structure or on process and will be discussed in subsequent chapters.

North and Teversham (1984) produced a detailed historical map of the riparian vegetation of three rivers in British Columbia. This effort provides a basis for landscape studies. The longitudinal pattern over many kilometers and the transverse pattern across 0–10 km is explicitly displayed. Tesserae as small as 2 ha are mapped. Thus aspects of both the longitudinal and transverse landscape structure and the internal structure are presented. They discussed both longitudinal and transverse gradients, and related these to geomorphic and hydrologic factors. Swanson *et al.* (1982) noted that studies of riparian vegetation are a small proportion of ecological work in mountainous regions, but emphasize its importance in ecosystem functions. The number of studies has increased in the past 10 yr.

Riparian areas in mountainous regions present some difficulties and opportunities for landscape ecology. One area is in the distinct longitudinal gradient observed, as noted above in arid mountainous areas (Campbell 1970; Pase and Layser 1977) and also in other mountainous areas (Knopf 1985; Olson and Knopf 1988; Finch 1991). In some cases, the relation between riparian areas and their surroundings is clear, and the ecotones are well defined. Even notable disturbances that affect these ecotones can be spatially distinct and thus present opportunities for the study of several aspects of the principles of landscape ecology (Wissmar and Swanson 1990; cf. Malanson and Butler 1991). In other areas, however, riparian areas are not so distinct. Many coniferous trees are able to grow in bog-like conditions, and a riparian area may have a mix of tree and shrub physiognomies. The overall spatial pattern may also be complex given elevational changes, the unknown effects of mountain barriers, and nonequilibrium conditions caused by climatic change. One approach that may prove fruitful is the application of the concept of centrifugal organization (cf. Keddy and MacLellan 1990). The variety of disturbances and other environmental determinants of plant communities in a complex situation may be able to be arranged in mountain riparian environments as Keddy and MacLellan (1990) have done for lakeshores and wetlands. Nilsson and Wilson (1991) have in fact compared the gradients on non-riparian mountainsides and lakeshores. It will be interesting to see if spatial variables *per se*, i.e. isolation, shape, connectivity, etc., can be included as axes in the centrifugal model.

Mountain regions emphasize the linear isolation of riparian areas more than their place in a mosaic. The ways in which this pattern is affected by geomorphology might be developed usefully in landscape ecology.

Taiga and tundra (Subarctic and Tundra Divisions)

While the studies of riparian ecology in the subarctic are relatively few, they include some of the most direct examples of work on the longitudinal pattern of riparian vegetation. Klokk (1981) included some information on longitudinal pattern related to an elevation gradient in a floristic study in central Norway. Nilsson (1986; Nilsson *et al.* 1988, 1989) has described the vegetation along rivers in northern Sweden. Nilsson *et al.* (1989) found that distinct floristic gradient existed on two rivers examined, which they related to climatic gradients. Nilsson *et al.* (1991*c*) reported that for two large and two small rivers, the longitudinal trends in species richness were similar if the smaller rivers are considered as miniatures rather than as pieces of larger rivers. In all four cases the highest levels of species richness were found where the rivers begin to cut into lacustrine sediments of a higher sea-level period (c. 9200 BP). Nilsson *et al.* (1991*c*) noted that this result may be due to multiple causes, but the spatial intersection of landscapes which may represent different rates of adjustment to global environmental change in the Holocene should also be considered. Further north, Kalliola and Puhakka (1988) mapped the riparian vegetation among peatlands in northern Finland in order to consider factors of landscape ecology such as river sinuosity, floodplain breadth, area, and rates of change, including channel abandonment. Most of the analysis was devoted to internal structure, but they related landscape structure to river dynamics.

The floristic composition of these riparian areas is not dissimilar at the genus level to that of the mountain areas: dominance by *Populus* and *Salix* on young sites or unstable substrate, and an apparent succession to spruce or other conifer. Bailey (1980, p.10), in his description of the more detailed early USA ecoregions map, described vegetation of the Yukon forest province as 'dense white spruce-cottonwood-poplar forest on flood plains' and on uplands 'stands of white spruce grow near streams'. Lee and Hinckley (1982) reviewed information on riparian areas from a variety of general ecological surveys in Alaska, and provided a detailed floristic summary.

In the subarctic, the overall species diversity is less than that at low latitudes, but is at its greatest in riparian areas. Because of the limitations imposed by the overall species diversity and the general uniformity of the environment over great distances, the landscape structure of subarctic areas is relatively simple. Studies of the internal structure, however, indicate that as much of interest happens here as elsewhere, and some address spatial processes (e.g. Walker and Chapin 1986) and are discussed in subsequent chapters. Thus subarctic landscape ecology has proven to be relatively well advanced. Problems, such as extensive bogs, in boreal areas will need to be considered. These riparian areas are of great interest, and northward flowing rivers are becoming increasingly important in hydrological management plans.

It may be of particular interest in landscape ecology to consider the mosaic structure of riparian zones in the taiga and tundra regions in reference to their relatively low diversity and the importance of allogenic forces in controlling fluxes across the mosaic.

Landscape ecology

Contrasts among ecoregions also emphasize contrasts among approaches in landscape ecology. In the arid and semiarid regions and also in mountains, where the riparian zone is visually distinct, questions about the linear nature of the area seem to be most relevant. In areas of a broad mix of elements, especially in the rainforest but also in temperate hardwood forests and taiga and tundra, mosaic processes of flows across multiple boundaries seem more important. Combining both approaches will be most useful in the long run, and areas such as the central USA and western Europe seem to have a natural advantage as well as a solid basis in past work for future developments in landscape ecology.

Testing the principles of landscape ecology requires that the structure be quantified because the most fundamental principle of landscape ecology is that landscape structure does affect process. It is necessary to test for differences in processes, but specifically for differences related to quantified structural elements. Quantification of landscape structure first requires delineation or mapping. While it is clear that delineation will be easiest in drier areas where riparian elements are more distinct, it is also possible to map riparian elements elsewhere. Remote sensing provides a tool that has been used for mapping wetlands in many environments (e.g. Hewitt 1990; Butler *et al.* 1991), and combined with geographic information systems and digital elevation models can do so for riparian landscapes (see Chapter 7). Furthermore, the testing of principles of landscape ecology will proceed where fundamental aspects of ecology, i.e. taxonomy, community structure, and ecosystem functions, are already known. Most important, however, is matching the landscapes we know with the questions we wish to answer. Nilsson's (e.g. Nilsson *et al.* 1989) work in Sweden is an example of the importance of asking the right questions, as opposed to finding the questions that fit the environment that is easy to define.

A global gradient in riparian landscape structure exists. The gradient is from the extensive and difficult to delineate riparian elements to spatially limited and/or confined elements that have distinctive boundaries with the non-riparian matrix. In the former category are the extensive floodplain forests in the basins of the Amazon and Zaire Rivers. In the latter category are the narrow fringes of riparian woodland along streams in the deep canyons in the southwestern USA. In between are the spatially extensive riparian areas associated with other wetlands in taiga and the Mississippi River embayment, the broad riparian areas not very distinct from surrounding upland forest in humid broadleaf ecoregions, the prairie and savanna types, whose distinctiveness may in part be maintained by fire, and the mountain areas where the landscape structure is directly controlled by geology. This gradient reflects a gradient in available moisture. In the wettest areas – those with high precipitation or low evapotranspiration – riparian landscapes are indistinct. In dry areas, the river itself is a great contrast with the surroundings, the gradient of moisture away from the river is sharp, and boundaries are clear. Comparisons of processes in these areas, perhaps beginning with some spatially explicit theoretical models (see Chapter 7), may be worthwhile.

In the USA, the Mississippi River system, including its major affluents the Missouri and Ohio Rivers, includes 19 ecoregion provinces. In this area latitude and longitude represent a climatic gradients of temperature and moisture. The variables have been identified over short transects (Lindsey *et al.* 1961) and where biogeographic discontinuities have been recognized (Robertson *et al.* 1978), but an overall assessment of landscape patterns has not been attempted. This diversity of ecoregions that combine to form the landscape structure of an entire river system is repeated throughout the world. This scale of landscape structure, where the riparian elements change from subtropical floodplain forests, to woodland–grassland transition, to true prairie and scrub steppe, and finally to subalpine and alpine source areas, presents a challenge in biogeography beyond the normal scale of landscape ecology, but one in which the principles of landscape ecology will apply.

The longitudinal gradient of landscape structure has in general received little attention in ecological research. This shortcoming has seen some work in the recent past, but the work in aquatic ecology, specifically in the River Continuum Concept (RCC) (Vannote *et al.* 1980), may provide a model for future development. This concept proposes a continuum in the abundance and activity of consumer organisms (shredders, grazers, collectors and predators) that varies with stream size because of differences in the inputs of coarse and fine particulate organic matter from both the upstream reach of the river and from the adjacent riparian zone. Field studies have all but rejected the concept because of the variability found in nature (e.g. Huryn and Wallace 1987) (in part due to the nonequilibrium geomorphic processes; the RCC began with the assumption of geomorphic equilibrium), but these results have led instead to interesting qualifications of the RCC and to further research. The relation of the changing form of the hydrograph to the longitudinal structure of the vegetation should be one aspect of this research (e.g. Gazelle 1989)

Studies on one topic might help to synthesize some aspects of the diverse landscape structure of the riparian environment of different regions. In many of the northern hemisphere riparian areas the genus *Salix* is of great importance. It dominates the early successional stages in the humid broadleaf forest and in the forest–grassland transition and grassland areas, and it may be the only specifically riparian woody genus in some mountain sites. *Salix* is found in the riparian zones in North America, Europe, and Asia. The biogeography of this genus needs to be studied. An evolutionary account of its development and geographical spread, and its response to changing climates, geomorphology, and river patterns and flows could provide major insights into riparian ecology at the landscape, regional and global scales.

In terms of the specific structural features identified by Forman and Godron (1986) for corridors, longitudinal gradients of height and width need to be investigated. These features can be related to regional trends in productivity and biodiversity. Forman and Godron (1986) placed great emphasis on the genesis of landscape elements, and the landscape structure of these processes of genesis needs further study. In riparian areas patch generation, such as the formation of a point bar, has particular spatial constraints. The spatial pattern of the geomorphic process will need to be investigated in combination with the ecological development on the patch because the two are correlated. Landscape structure seems to be a potentially fertile area for investigation. It will be seen to affect landscape processes, and a solid base of ecological research on the internal structure of riparian areas provides a good basis for further work.

Conclusions

The riparian environment differs in its landscape role among ecoregions. A general conceptual gradient from a strongly linear landscape feature to a part of a more complex mosaic can be seen by contrasting ecoregions. The differences in landscape structure are emphasized in this view. While this gradient reflects our visual perception, it is also related to the abruptness of the terrestrial–aquatic ecotone. The differences in hydrology and vegetation also differ among regions and the complexity of ecological processes is clear in the contrasts.

4 · *Internal structure*

The number and constant successions of these islands, all green and richly wooded; their fluctuating sizes, some so large that for half an hour together one among them will appear as the opposite bank of the river, and some so small that they are mere dimples on its broad bosom; their infinite variety of shapes; and the numberless combinations of beautiful forms which trees growing on them present: all form a picture fraught with uncommon interest and pleasure. *(Charles Dickens)*

The riparian landscape is heterogeneous, i.e. it contains an assortment of tesserae whose configuration is itself important. The arrangement of tesserae and the assemblages of species comprise the internal structure of riparian landscape elements. In reference to vegetation, Rejmanek (1977) defined structure in three steps: the identity of species (presence–absence), the relative abundances of the species, and their spatial pattern. If we consider step 3 to be the arrangement of communities, not simply individuals, then this definition can serve as a starting point for considering internal structure in riparian landscapes. The relative abundance and configuration of tesserae determine the internal heterogeneity of the riparian landscape element, and heterogeneity varies among regions and is related to regional and local controls (Baker 1989).

Basic descriptions of riparian environments are a starting point for consideration of internal structure. Early ecological reports provide an array of concepts, ranging from visual to structural classifications, which differentiate among tesserae. For example, Ridgway (1872), Perkins (1875), and Turner (1936) gave interesting descriptions of the variety of tesserae in riparian vegetation in Illinois. These descriptions emphasize the visual cataloging of multiple ecosystems within riparian zones. From another perspective, Tchou (1951) compiled extensive releve data on riparian vegetation in southern France. These efforts to describe, classify, and explain the heterogeneity within the riparian zone only hint at the broad topics that have been considered in studies of particular river reaches or even individual tesserae.

Community structure

I will consider three paradigms that are fundamental to many studies of riparian ecology. They are the individualistic hypothesis of plant associations, the intermediate disturbance hypothesis, and the competition hypothesis of niche relations. I will introduce these paradigms and then examine major topical approaches to riparian ecology.

Riparian habitats present a classic case of a clear environmental gradient. The basic definition presented in Chapter 2 referred to the gradient from upland terrestrial conditions to aquatic conditions. This gradient has been the subject of many studies in plant ecology. These studies seek to explain the distribution of plant species along this gradient or the relative abundance of species at specific points on it. The individualistic concept of species distributions on environmental gradients (Gleason 1926) is a dominant paradigm in ecology. While the universality of this concept is not certain (Westman 1983), and the form of the distributions of species on the gradient is contentious (Austin 1987), the general idea that the environmental distribution of species is not interdependent has been widely accepted, and gradient analysis as a method of investigation is widespread (e.g. Kupfer and Malanson 1992*b*). It is critical that the concept of an environmental gradient not be confounded with an actual transect laid out on the ground. The two may coincide in a few obvious places, such as the gradient of elevation as one moves up a smooth mountainside, but in many places a gradient in an environmental variable will not coincide with a transect. On floodplains, it is notable that distance from the river channel does not provide a gradient in most hydrological conditions, because the topography and hydrological history of the location are complex. Thus to construct a gradient in, say, depth to the water table on a floodplain, one must locate sites through a full two-dimensional array on the floodplain and then rearrange the order of sites according to depth.

A second paradigm is the intermediate disturbance hypothesis (Connell 1978). Species diversity may be greatest where disturbances are intermediate in frequency, area and intensity (cf. Malanson 1984), and each of these factors can be considered to be a separate environmental gradient. In riparian areas, this concept is thought to relate to flood frequency. The spatial nature of the disturbance and of species responses in riparian areas complicates this relation, however. It is important that the processes leading to alpha diversity, the number of species growing together on a site, not be confused with those leading to gamma diversity, the number of species in a region, although the same processes may affect both. The same basic methods of observation, inference, and hypothesis generation that are used for the individualistic hypothesis can be applied, for species diversity, to the intermediate disturbance hypothesis. More focus is given to life history traits than to physiological mechanisms in this area.

A third paradigm is that of the importance of competition. Competition may presently operate in ordering a hierarchy of species, or it may have operated in the past to produce distinct niches. In one dimension, the niche is the distribution of a species on an environmental gradient, and a niche hyperspace is the niche expressed mathematically on *n* gradients. The concept that competition to some degree limits species responses to the environment, and that release from competition would result in species expanding their niche, underlies much of the work in riparian ecology related to environmental gradients and disturbance. Keddy (1989) explained two hypotheses that can account for the common pattern of the distribution of species along environmental gradients. One is that through competition in the past, species have adapted to the conditions of their environment quite closely, so that their range of tolerances and requirements is close to the environment in which they are actually found. If competition operated in the past to control species niche widths, then present realized niches should be quite similar to fundamental niches. The second is that a competitive hierarchy exists, and that some species could exist in environments beyond their present range, specifically in the direction of less stress, except for the presence of competitors. If a competitive hierarchy is operative, then the present realized niches of some species are much smaller than their fundamental niches. Keddy (1989) wrote that the way to distinguish between these two hypotheses was through removal experiments. He and colleagues have done several such experiments on lake shore environments, which share some similarities with riparian environments, and found clear evidence for the competitive hierarchy. The basic idea behind the competitive hierarchy is not new: Darwin (1859) noted that on a climatic gradient many species can grow beyond their natural range in gardens, and MacArthur (1972) wrote that the southern range limits of species would be limited by competition, not climate.

Although in some disrepair, the model of Grime (1979) expressing a simplified notion of a hyperspace made up of a disturbance axis and a productivity axis, which gives three extremes dominated by low productivity and stress, high productivity and competition, and disturbance, is a good simplification (the differences between this concept and that of Tilman (1988) being largely semantic (Grace 1991)). Any line in the triangular model is an environmental gradient; the disturbance gradient is perpendicular to the productivity gradient and this added dimension allows more species to coexist; the place of species on the diagram depends on the locations (in time and space as well as on the figure) of other species.

Riparian gradients

Major environmental gradients are considered in examining the internal structure of riparian plant communities. The gradients within local areas are related to fluvial dynamics, floods, and soil moisture. These three factors are related. Channel dynamics create the topography of the riparian zone, and in turn are controlled by the topography. The topography directly controls flooding, and the frequency and duration of given discharges, and river stages, will have definite spatial distributions in any riparian area: low areas are flooded more often and longer. These low areas can also extend flood duration by trapping water in surface storage. Still waters lead to the deposition of fine sediments, and these topographic depressions thus have very low infiltration rates, which increase surface storage and flood duration. The surface storage on fine sediments provides a longer source of influent water to groundwater and thus a higher water table is maintained. These sediments are often saturated. Topographic ridges, conversely, are flooded less frequently and for shorter duration. Over-bank flows run off quickly and the sediment deposited is relatively coarse. Infiltration and drainage is rapid, the groundwater table lower, and the sites relatively dry. These differences can occur with topographic variations of a meter or less on a single terrace level, and where distinct terraces are found the effects are multiplied. Channel dynamics, flooding, and soil moisture are all affected by hydraulic control structures. I will discuss studies that identify themselves with these three major ways in which gradients have been considered: channel dynamics, soil moisture, and generalized flooding, even where the actual mechanism in operation is the same, e.g. anoxia. Within the topic of soil moisture I will examine experimental studies on anoxia, and within flooding I will examine the effects of dams, diversions, and channel alteration.

These gradients can be generalized as a transverse gradient perpendicular to the channel. The actual transverse pattern found in riparian areas is much more complicated than a single environmental gradient that spatially coincides with a transect from the river to the upland (Figure 4.1). Topographic variation is the common element. The patterns that

Figure 4.1 A cross section of the riparian zone indicates the topographic diversity and its connection to the water table.

are seen in relation to the generalized environmental gradient will be examined first, because most studies have emphasized this approach, and then their complex spatial configuration will be considered. Ware and Penfound (1949) found that vegetation patterns were controlled by a combination of factors including shifting sands, mechanical destruction by flood waters, and lack of soil moisture. Aspects of these processes, and moreover of excess soil moisture, are seen in most studies of the internal structure of riparian vegetation. While the topics discussed below are distinct, some studies have considered many of them directly or indirectly (e.g. Wistendahl 1958). Sollers (1973) related vegetation patterns to soil type, which in turn was related to former channel dynamics. Johnson and Vogel (1966) described vegetation mosaics in the riparian vegetation of the Yukon flats; even in this time before satellite imagery, they noted the usefulness of examining such mosaics using multiple imaging sensors. A gradient of microclimate, in part related to light penetration and soil moisture, probably exists within the riparian zone but has not been quantified; it, too, may have some general relation to topography. Other factors such as fire, grazing, or permafrost may alter these relations (Bren and Gibbs 1986; Crampton 1987).

Welcomme (1979) gave some interesting information about internal structure of riparian floodplains in general and for specific cases. Following Leopold *et al.* (1964), he identified seven geomorphological features which bear repeating here: abandoned meanders or oxbows, point bars, meander scrolls or scroll bars, sloughs associated with scroll bars or at floodplain margins, levees at cut banks, backswamps, and sand splays. He provided a generalized map of how these features might be arranged. He also mapped the floodplains of five rivers and two deltas in the tropics. For these he only showed permanent and seasonal flooded areas. For a number of African rivers he provided data on the areal extent of the river and channels, lagoons, and swamps. From this report, meant to introduce the diversity of habitat associated with floodplain fisheries, it is clear that the environmental gradients related to moisture, soils, erosion, deposition, and channel dynamics will not be directly correlated with simple transects or cross sections of the riparian element.

Channel dynamics

Studies of the effects of channel dynamics on forest structure are most common in alluvial rivers where channel position changes readily and where scars of earlier meanders leave a record of geographical instability (e.g. Schumm and Meyer 1979; Gazelle 1987; Marston 1992). The spatially transgressive nature of this phenomenon may have important implications for landscape ecology. Shull (1922) recognized this phenomenon in observations of the growth and vegetative colonization of an island in a channel of the Mississippi River. Hefley (1937) claimed that it was his boyhood familiarity with such changes along the Canadian River in Oklahoma that led him to his research. This early study emphasized the deposition of sand in over-bank floods. Kalliola and Puhakka (1988) emphasized the topographic diversity of floodplains in both their specific case and for riparian ecology in general. Hughes (1988) noted that, for arid and semiarid African floodplains in general, minor topographical variations had an important effect on species distributions through their role in determining the flood regime. She detailed this analysis for her own study site on the Tana River (Hughes 1990). She specifically described the landforms and associated species of active and inactive levees, point bars and scroll bars, and abandoned channels. Her maps and cross sections of the internal structure of vegetation are an excellent example of the relation between channel dynamics, flooding, sedimentation, and vegetation.

Many of the geomorphological features of the riparian zone are thus ephemeral and the species that inhabit them must be fugitives (*sensu* Hutchinson 1951). Decamps *et al.* (1988) considered the internal structure of the riparian landscape of the Garonne River in France. They presented a model of riparian forest dynamics in which they argued that cyclical or 'reversible' dynamics could occur within certain parts of the riparian zone which were subjected to repeated flooding, erosion, and deposition, but that higher terraces could develop where the flooding cycle was broken, dynamics were no longer reversible, and autogenic forces dominated. They noted that human activities have increased the proportion of riparian land in the latter category by regulating flow.

This study is a good example of the consideration of the internal structure of the riparian zone from a landscape perspective.

The vegetational processes associated with these dynamics have been considered in terms of a vegetation type maintained by disturbance and/or as some form of succession. Succession is a directional change in species composition. It may be driven by either autogenic or allogenic processes. If the processes change, the rate and direction of change in species composition will also. The earliest studies of vegetation dynamics on riparian sites did not refer to the concept of succession (Fitzpatrick and Fitzpatrick 1902), but nevertheless illustrated the process. Many of the successional processes in riparian areas can be classified as primary succession: the establishment and development of a new plant community in an area without the effects of a previous plant community in that location. Most succession in riparian areas arise where new terrestrial sites are created by channel dynamics. These sites can be classified into two groups: first, depositional sites (mid-channel bars in braided rivers, point bars, and older terraces), and second, abandoned channels.

In the following sections I will discuss the observations made on the internal structure of riparian areas responding to channel dynamics in the context which the studies themselves have emphasized: succession, including disturbance and establishment, phenology, and temporal sequences; channel abandonment; and edge effects. These categories reflect key aspects of the competition-gradient approach in plant community ecology.

Successional studies

Disturbance and establishment

Throughout riparian areas, from the alpine sources where steep gradients and paraglacial sediments coexist, to the great deltas, the flux of sedimentary deposits is such that topographic sites available for the establishment of vegetation are constantly created and destroyed. Sediment moves in pulses controlled by hydrological events and a significant portion of the sediment in an active system is stored in channel and floodplain deposits at or near the level of the river (Kelsey *et al.* 1987). In these spatially dynamic environments many species coexist regionally by occupying different stages in the successional sequence, but even within the initial stage, distinct differences in vegetation are detectable among stands. Because such stands are young and small, stochastic processes of

Figure 4.2 Gravel bars in montane rivers can accumulate large amounts of coarse woody debris; Middle Fork of the Flathead River, Montana.

dispersal and founder effects may be important. These responses may be of importance to biological diversity and to the effects of human impacts.

Channel dynamics lead to obvious disturbances in mountain environments where the sediments are coarse. In Canada, Nanson and Beach (1977) concentrated on active geomorphic processes of rivers. They reported the close association between the geomorphic process of point bar formation and vegetative succession. Teversham and Slaymaker (1976), North and Teversham (1984), Bradley and Smith (1986), and Fyles and Bell (1986) have focused on specific examples of primary succession on newly deposited material in and near the Rocky Mountains in Canada. These studies again emphasize the deposition of fine sediment in environments where large clasts and gravel bars are active geomorphic features.

Malanson and Butler (1990, 1991) examined the role of large woody debris in affecting the relation between deposits of fine sediment and vegetative succession on relatively recent gravel bars in braided and anastomosed sections of the Middle Fork of the Flathead River, Montana (Figure 4.2). We found that hypotheses of a positive effect of woody debris on the deposition of fine sediment was not supported, while woody debris and the incorporation of organic matter in the soil and the vegetation composition and diversity was related, but that an overall system of relations describing the combined effects of woody debris, topography, and sedimentation on vegetation could not be supported. Temporal or founder effects on these recent sites, their small size, and the effects of paraglacial sedimentary processes were considered to account for the lack of clear relations (Shaw 1976; Lee 1983; Lee *et al.* 1985; Jenkins and Wright 1987).

In California, McBride and Strahan (1984*a*,*b*) examined the role of channel dynamics in the establishment and survival of seedlings, and Harris (1987) used gradient analyses to study patterns of species occurrences in relation to channel dynamics. Campbell and Green (1968) described relations between succession and sediment in Arizona. Bellah and Hulbert (1974) in Kansas and Sollers (1974) in Michigan also studied dynamic processes, sediment, and vegetation. These studies emphasize the importance of substrate conditions for the early stages of succession. Many of these studies of establishment in riparian areas have involved *Populus* species. From the early work of Shull (1922) and McVaugh (1947) up to current studies (Baker 1990*b*; Malanson and Butler 1991) this genus, along with *Salix*, has been perceived as of unparalleled importance in this context.

In detailed studies of the establishment stage in Alaska, Walker *et al.* (1986) reported experiments on the conditions in which species seeds can germinate, and Walker and Chapin (1986) conducted experiments on the nutrient, water, and competitive relations among successional species on an Alaskan floodplain. The successional processes are controlled by site conditions related to flooding and sediment and by autogenic processes related to competition. In a study also very focused on the establishment phase, Conchou and Pautou (1987) found that differences in the initial colonization and subsequent growth form of a grass species were related to sediment type, flood durations, and fertility.

Baker (1988) focused on size classes as affected by flood events along the Animas River, Colorado. He found that a single flood event, and its lasting effects, may have changed the course of succession. He noted that repeated flood events, rather than a regular successional sequence, may explain the patterns of vegetation found along the river. Disturbance, which alters the conditions for establishment, is also important in maintaining diversity (Baker 1990*a*). In a study of the factors affecting the dominant species in this environment, Baker (1990*b*) reported that climatic effects determine when conditions for seedling establishment are propitious, and that stands originate in these years more commonly. The frequency of such years may be increasing as climate changes.

Flooding may favor some species by preparing a seed bed in lowland areas also. Hosner and Minckler (1963) noted that flooding provided the bare mineral soil seed beds needed by some riparian species. The temporal development of vegetation along the Missouri River in the prairies of the USA in North Dakota has been described by Johnson *et al.* (1976). They differentiated among pioneer and successional species, and found that pioneers established on newly exposed bars, and were succeeded by other species when these places became more isolated from active river processes as terraces. In this environment, *Populus deltoides* and *Salix amygdaloides* dominate the pioneer stage, and are replaced by *Acer negundo*, *Fraxinus pennsylvanica*, *Quercus macrocarpa* and *Ulmus americana*. Johnson *et al.* (1976) considered the soil surface to be of particular importance in the establishment phase, and recorded its change through succession.

In California, Laymon (1984) observed rapid riparian vegetation development subsequent to channel shifts. Lisle (1989) described a feedback between riparian trees establishing at low water levels which may then protect and define a new streambank in a laterally mobile stretch of river. Trush *et al.* (1989) distinguished riparian plant communities associated with active channel and 'bank-full' (bench or channel shelf) internal zones. The tree densities in the channel zone were limited by a threshold of channel gradient and curvature, while those in the bank-full zone were unrelated to these measures. Barnes (1985) examined the population dynamics of several riparian tree species in Wisconsin on a mid-channel island that had been accreting, primarily by deposition on the upstream end, for at least 44 yr. He reported differences in the development of vegetation because of differences in the sprouting and clonal development of species following damage by flooding, and noted the role of beaver (*Castor canadensis*) in controlling the development of some species (see below).

The work in the Peruvian Amazon by Salo *et al.* (1986) also emphasized channel dynamics in relation to a regional pattern of plant diversity. Campbell and Frailey (1984), Salo *et al.* (1986), Rasanen *et al.* (1987), and Frost and Miller (1987) have considered that the geological and related geomorphological processes of river dynamics have contributed to the development of species diversity in the Amazon basin. They argue that river processes continually create new landforms and thus expand the niches available for species. This argument accords with current theory on the role of disturbance in promoting biological diversity (Grubb 1977; Connell 1978). A number of authors have considered the successional status of species in tropical riparian environments (Fanshawe 1954; Bacon 1990, for Guiana). Paijmans (1990) illustrated major changes in river form for the Musa River in New Guinea, and described several successional sequences, noting that these dynamics are most obvious where land accretion is rapid. Lamotte (1990) specifically described relations between fluvial dynamics in the Peruvian Amazon; she identified one grassy and two shrubby pioneer stages. Although the argument for the role of river dynamics in promoting tropical diversity is attractive, Dumont *et al.* (1990) have questioned them, because upland sites are in fact more diverse, but this analysis does not reject the hypothesis in relation to the specific process. Salo (1990) provided one of the most explicit descriptions of the application of landscape ecology concepts to a riparian environment. He emphasized the relation between geological and geomorphological processes which create the arrangement of channels, floodplains, abandoned channels, and erosional and depositional sites, and he noted that ecotones are spatially dynamic in this environment; he did not, however, discuss the spatial configuration of the landscape elements in detail. The spatial pattern of the primarily depositional sites and the importance of founder effects leads to the idea that species movements among tesserae are important components of the reproduction of these landscapes (also see Chapter 6). Meave *et al.* (1991) have also argued that riparian forests could have served as Pleistocene refugia in Venezuela and Belize. When the disturbances are anthropogenic, the resulting patterns of establishment may be uniquely related to the abiotic environment and provide greater beta diversity (cf. Morin *et al.* 1989).

Phenology

Dispersal or germination phenology may also affect the assemblage of species and their relative success in particular riparian environments. The timing of dispersal relative to the hydroperiod is important. Zimmerman (1969) noted that riparian trees in Arizona (*Platanus wrightii*, *Populus fremontii*, *Salix* spp.) have dispersal phenologies similar to their eastern congeners, not to their upland neighbors. With dispersal, seedlings are established on wet substrates following spring floods, rather than following summer rains. Noble (1979) reported that differences in the establishment of *Populus deltoides* and *Salix interior* on point bar environments in the midwestern USA occur based on the intersection of flood timing and their timing of seed release. Currier (1982) did not find distinct differences in the period of seed release of *P. deltoides* and *S. exigua* and *S. amygdaloides*, but he did note that it was necessary for exposed mudflats to be accessible during the viability period, which he identified as mid-May to August. This study was based on 71 1 m^2 quadrats in which only 26 seedlings were recorded. This period is longer than I would estimate based on 106 quadrats with seedling densities ranging from 0 to 595 (Craig 1992). The period of viable seed release certainly varies geographically. Seeds which disperse at just the right time when floods have peaked and are receding will find the needed moist seedbed at such sites (Craig 1992). Since the timing of floods will vary, each species has different years in which it is successful, and so competition is, in theory, reduced. Malanson and Butler (1991) speculated that this same effect operated on gravel bars in Montana. Otherwise similar channel bars had nearly complete monospecific dominance by either *Populus* or *Salix*. Since dispersal phenology and flood timing vary here too, we hypothesized that this coincidence of events might lead to the differentiation of dominance in this environment as well.

McLeod and McPherson (1973) found soil moisture limiting for the establishment of *Salix nigra*. Bradley and Smith (1986) also related river stage during the period of seed release by *Populus deltoides* to their successful establishment, and considered the disruption of this co-occurrence the cause of their decline below a dam in Montana. Howe and Knopf (1991) reported that a change in flooding has favored the invading *Eleagnus angustifolia* and *Tamarix chinensis* over the native *Populus fremontii* along the Rio Grande in New Mexico. Niyama (1990) studied the effect of dispersal phenology on the coexistence of six riparian *Salix* species in Japan. He found that a distinct gradient in the timing of seed release and preference for soil texture, and degrees of specialization on the gradient, led to distinct species associations, which he attributed to the regeneration niche. *Populus* and *Salix* species are notable as riparian pioneers worldwide, and this aspect of their biogeography would be a good starting point for a comparative study.

Extensive studies have been made on the establishment of *Tamarix* in the southwestern USA (e.g. Horton *et al.* 1960; Robinson, 1965; Harris 1966; Turner 1974; Warren and Turner 1975; Graf 1978, 1982). These studies have concluded that plentiful seed dispersal, phenology that is well suited to the flow regimes of the region, and human impacts affecting the sediment budgets of the rivers of the region have contributed to the spread of *Tamarix* and its growth in dense stands, which

Figure 4.3 Newly establishing *Tamarix* increases the deposition of fine sediment; Colorado River, Utah.

increase the deposition of the sediment in which it thrives (Figure 4.3).

Jones and Sharitz (1989) contrasted the effects of germinating times of floodplain trees keyed to avoid canopy shading and flooding, but this laboratory trial did not specifically address the actual timing of events that might occur in the field. Streng *et al.* (1989), in contrast, found that specific events such as flooding had a significant effect on the seedling survival of other floodplain trees, and related the timing of such events to the timing of dispersal and germination.

Temporal sequences

In respect to channel dynamics in the southwestern USA, while Campbell and Green (1968) described the process as 'perpetual succession', Johnson *et al.* (1989) denied that succession actually happened in this region. The latter authors noted that only one type of plant community exists along the rivers, and that when it is destroyed by channel dynamics or when new sites are created, this community establishes and is never replaced by a different one. Although there may be some variation in the relative abundances of species in the establishment phase, if these are due only to founder effects then there truly is not succession as strictly defined. Johnson *et al.* (1989) may or may not be focused too narrowly, however. In some cases the channel dynamics are such that an area once riparian is left farther and farther from the river, and in these cases the riparian vegetation is replaced by the dominant desert scrub. Because Johnson *et al.* (1989) focused on the question of whether or not succession occurs within riparian areas *per se*, they may not have taken this coarser-scale phenomenon into account. However, in many areas, where the river is confined in a canyon, this phenomenon may in fact be rare and their analysis would stand.

Irvine and West (1979) described successional stages related to flood terrace heights along the Escalante River in southern Utah in an area where the floodplain occupies the bottom of a canyon about 100–300 m wide. In another arid environment, Halwagy (1963) described vegetative stages on deposits of the Nile. The ages of some deposits were known well enough to distinguish among seres, and autogenic processes could be inferred.

Successional processes on depositional features have been examined in several other environments. In temperate environments, emphasis has been on the alluvial features, sediment deposits, and soil development (e.g. Shull 1944). Wilson (1970) and Sedgwick and Knopf (1989) both described what seem to be a classical cases of facilitation involving increasing deposition. In subarctic and mountain regions the focus has been on the geomorphic processes in the channel, soil temperature and permafrost, distinct geomorphic terraces related to fluvioglacial activity, and to these and soil development (e.g. Drury 1956). Gill (1972*a*, 1973) described a sequence of successional seres related to specific depositional levels, including new sediments, prograding slopes, terraces, and meander scrolls (cf. Farjon and Bogaers 1985). Anderson (1947) briefly described the stands of *Salix* and *Platanus* on fresh gravel bars in Missouri as a situation with little competition, thus revealing founder effects. Notable feedbacks between channel dynamics and species establishment and growth are also probable. Dietz (1952) described the relation between the establishment and growth of a gravel bar in Missouri and the establishment and growth of *Salix* spp. on it. He noted that following floods, lines of willows establish along the waters edge and these then serve to advance the edge of the bar by resisting floods and trapping sediment (Figure 4.4).

Brief reports from Oklahoma (Burns 1940; Featherly 1940) found successional trends and a major change in species composition related to rapid sedimentation which led to poorly drained conditions. The

Figure 4.4 *Salix* species can establish in narrow bands at the river's edge, depending on soil moisture and accessibility; Cedar River, Iowa.

dynamics of the Mississippi River were the focus of a detailed floristic study by Shelford (1954) in the Tennessee reach. He used detailed maps of channels, extending back to c. 1500 BP and of 10 plant communities to relate vegetation to fluvial processes. He also reported specific responses to flooding, distinguishing between erosive and depositional events and among sediments deposited. Shelford (1954) distinguished two major vegetation groups related to short and long duration flooding which he then attributed to differences in alluviation. He also documented wildlife and aquatic habitats.

In Alaska, Bliss and Cantlon (1957) sampled a cross section of the Colville River, Alaska. They described four vegetation groups arranged in temporal sequence associated with the migration of a river meander. They related the differences among the plant communities to changes in elevation above the river, flooding, and sediment deposition. These general relations, i.e. that the deposition of fine sediment increases with vegetation growth, leading to higher site elevation and different conditions for vegetative growth, are those that have been documented in more detail in subsequent studies. For subarctic studies, they also noted the changes in soil temperatures that occur during succession. Viereck (1970) illustrated a fuller development of the successional sequence in a study along the Chena River, Alaska. He examined differences among stands of 15, 50, 120 and 220 years of age, and presented a more detailed examination of the role of soil temperature in this region. Crampton (1987) also described the temporal change in permafrost related to the migration of a meander and the succession of vegetation on the migrating point bar. Lee and Hinckley (1982) summarized information pertinent to riparian vegetation from a variety of reports, mostly on general vegetation and ecology. They reported domination by *Salix* and *Populus* species that varied with substrate and soil related to the dynamics of rivers creating terraces and channel features. Farjon and Bogaers (1985) inferred seres from geomorphological patterns, especially meander features.

Fonda (1974) described older successional sequences on river terraces associated with glacial stages in the Olympic Mountains, Washington. Hawk and Zobel (1974) also addressed soil development on alluvial landforms in this region. Shaw (1976) and Lee (1979, 1983) documented different plant communities at different stages of succession in the Pacific Northwest, and Jenkins and Wright (1987) modeled the succession of riparian vegetation and related successional sequences to wildlife habitat. In a mountain setting in Norway, Klokk (1981) diagrammed successional pathways of riparian vegetation. He identified seven distinct initial communities (e.g. *Salix nigricans–Lotus corniculatus* on gravels and *Carex juncella–Salix glauca* on permanently moist backwaters) and two climax types (birch and spruce forests).

A comprehensive study of riparian vegetation is that of Lindsey *et al.* (1961). They examined fluvial processes, soils, climate, and flooding. They identified eight physiographic categories of the floodplain and seven successional seres on it. Phytosociological analyses in the floodplain forest showed a continuum along the 370 km latitudinal gradient. Thus the authors identified two major factors affecting the maintenance of plant communities: recurrent fluvial dynamics and regional climate. They did not discuss their interaction nor climate dynamics. A related gradient analysis of riparian forest in Ohio revealed patterns related to geomorphic stand age and secondarily to degree of human disturbance (Hardin *et al.* 1989).

From another perspective, Dahlskog (1966) related depositional processes to flooding, topography, and vegetation on a lacustrine delta in Swedish Lapland. Reversing this perspective, Waldemarson Jensen (1979) wrote of the numerous different pathways of succession engendered by the development of another delta in a lake in Sweden. She used aerial photos and both present vegetation and vegetation excavated from recent sediments to derive a very complex model of alternative pathways. She used a classical paradigm of succession and identified progressive, retrogressive, and topographical climax conditions. Sedimentation, flooding, and the accumulation of dead plant material were identified as important components of the process. Because a variety of pathways are identified specifically in spatial relation to alternative pathways which are well defined because of the geomorphological controls, i.e. of levees and lagoons, her study presents an interesting look at complex arrangements of tesserae and their interaction within the riparian landscape element. Other studies of lake deltas have also found complex patterns of vegetation associated with channel dynamics (Dirschl and Coupland 1972; Butler *et al.* 1991).

M. Izard (personal communication) has found that successional sequences differ along the longitudinal profile of the Tech in southwestern France, and several studies along the Rhone have documented successional processes. Some of these studies have been concerned primarily with a phytosociological description of the vegetation (e.g. Pautou 1980; Pautou and Girel 1980, 1982, 1988; Girel and Pautou 1982), while others have been more closely related to studies of the dynamic geomorphology of the river (e.g. Bravard 1986, 1989, 1990). Pautou (1984) and Pautou and Decamps (1985) noted that the combination of allogenic forces of hydraulics and sedimentation combined with autogenic processes related to changing soil and hydrology led to a spatially heterogeneous vegetation; in other words, the internal structure is complex. Pautou *et al.* (1985) and Bravard *et al.* (1986) used historical information about hydraulic structures to develop a model to describe several related successional sequences that would occur on different parts of the transverse structure of a braided and anastomosed floodplain. They found that the speed of succession was controlled by the balance of autogenic and allogenic processes, which were affected by their topographic position and geomorphic history. These studies place much more emphasis on the landscape-level hydrological processes than do those concerned with on site sedimentation. Amoros *et al.* (1987*a*,*b*) took these studies and developed their significance to environmental management by examining how water management might affect successional processes and serve to continue some, recreate others, or eliminate stages altogether. Bournaud and Amoros (1984) considered the role of biological indicators for management in this environment. Also in France, Decamps *et al.* (1988) have proposed a model similar to that of Amoros *et*

Figure 4.5 Abandoned channels remain as topographic lows in the riparian zone; White Breast River, Iowa.

al. (1987*b*), but in which at some stages seres can become isolated from the riverine processes and no longer part of the normal shifting mosaic.

In most of these studies of succession allogenic forces are very important. This fact should alert us to the landscape level processes that are active. By their very definition, allogenic forces are ones coming from the landscape to an individual tessera.

Channel abandonment

The topographic forms imposed by abandoned channels are important factors in the differentiation of the internal structure of riparian landscapes (Gesink *et al.* 1970) (Figures 4.5 and 4.1). The process of channel abandonment occurs on the wide floodplains of major alluvial rivers. Chapman *et al.* (1982), while summarizing information about wetland and riparian species' tolerance to flooding, presented diagrams of the internal structure of floodplains from several regions which have representations of abandoned channels. Shelford (1954) included abandoned channels in his maps of riparian vegetation along the Mississippi River. Where meander bends are extreme, the river can cut off sections of itself, forming oxbow lakes. In braided and anastomosed areas, the several channels can shift, and some be abandoned. Abandoned channels subsequently dry and fill with sediment. Their former banks, however, often remain higher. Alexander and Prior (1971) found differences in rates of sediment deposition on ridges and swales of the floodplain of the Ohio River to be small and dependent on location over 10 000 yr. These ridges and swales continue to control the flows of subsequent over-bank flooding. Abandoned channels become temporarily reactivated; even when an entire floodplain is inundated, these channels will route more flow. These channels also hold water after flooding or intense precipitation. In Iowa I have observed abandoned channels to hold surface water from spring floods for most of the summer growing season. These channel features are closer to the water table, and it in fact may be nearer the surface because of them: when these areas hold surface water they are probably an influent source of water to groundwater that is even lower below the nearby ridges. An associated factor is that the sediment in the channels is much finer than that of the ridges. This fine sediment may act to retard infiltration, maintain surface storage, and increase anoxic conditions. Abandoned channels at different stages then represent an internal structure which affects ecological processes through their effect on subsequent channel dynamics (M. Izard, personal communication), flooding, and soil moisture. Here I discuss the former, and the other two will take their place in later sections of this chapter.

The most notable body of work on abandoned channels has been done along the Rhone (Figure 4.6). The original studies were under the direction of P. Ozenda at the University of Grenoble (Ain *et al.* 1973) and have since expanded to include researchers at Lyon. Castella and Amoros (1984, 1986) described the plant communities of abandoned channels or 'bras-morts' (dead arms). They found associations with the origin of the water and geomorphological features, and considered both allogenic and autogenic processes in the vegetation change in these channels. This study was followed by studies of the macroinvertebrates in these abandoned channels, which can serve as indicators of the successional stage, in these cases from aquatic toward terrestrial environments (Castella and Amoros 1984, 1986, 1988; Amoros and Jacquet 1986; Castella *et al.* 1986; Richoux and Castella 1986; Richardot-Goulet *et al.* 1987). These studies found that certain invertebrates were useful indicator species related to specific hydrological processes, and that they could be used to reconstruct the history of environmental change in some abandoned meanders (Green (1972) also examined invertebrates in abandoned channels in Brazil). Rostan *et al.* (1987) characterized aban-

Figure 4.6 Aerial view of an abandoned channel on the Rhone (courtesy J.P. Bravard).

doned channels using ratios of organic matter (defined as C × N) and $CaCO_3$ in bottom sediments to differentiate among abandoned channels. They found differences between those channels that had been abandoned meanders or anastomosed channels which develop a terrestrial succession dominated by *Alnus glutinosa*, and those that had been braided channels and develop a succession dominated by *Ulmus campestris*. Other work has focused on the geochemistry of these areas and their soil fauna (Dole 1983, 1984; Carrel and Juget 1987).

Gastaldo *et al.* (1989) reported patterns in the preservation and diversity of biological detritus in an abandoned channel oxbow lake in Alabama. Differences in the materials were found between levee and channel bottom areas, which depend on both differences in input and differences in preservation. In Iowa we have examined the sediment texture and chemistry of abandoned channels along the Cedar and Iowa rivers in relation to their position in the landscape relative to the current river channel and to agricultural fields and to time since abandonment (W. Schwarz, personal communication). Phosphorus, nitrogen, and potassium were not associated with connectivity to the active river or with time since abandonment, but were negatively associated with proximity to agricultural fields. This result is counter-intuitive to hypotheses based on the principles of landscape ecology. The explanation is in the differences in the very local hydrological conditions wherein geochemical transformations and cycles differ with the amount of water stored at the site. The local effects of hydrology, especially when standing water is present, have been demonstrated in more detail by Forsberg *et al.* (1988), who found a variety of amounts of nitrogen and phosphorus in lakes of the Amazon floodplain.

Shankman and Drake (1990) found that *Taxodium distichum* may be dependent on abandoned channels in the northern part of the Mississippi embayment. They found that it colonizes the shores of abandoned oxbow lakes and then the lake itself as it fills with sediment. This species can dominate these sites for centuries, but cannot reproduce under a closed canopy, and so seems to be dependent on continual channel dynamics of this kind. Given the differences in the initial development of these features, this later distinction in major forest cover type is expected. Farther north, *Acer saccharinum* and *Betula nigra* may be in a similar position relative to abandoned channels. These species always occupy the lowest slough areas of a floodplain or the edge of abandoned channels which regularly hold water. *B. nigra* is here in a precarious position, being near the northwest corner of its range limit, and is usually found in poor condition. Many abandoned channel sites have dead *B. nigra* along their edges, and the only extensive regeneration that I have observed is along the edge of an abandoned channel downstream from a dam. *B. nigra* seems to be sensitive to high nutrient levels and may be successful next to flowing water but not in a less open system. Where they are thriving may be due to fewer nutrients in the river below the dam. Other explanations, such as anoxia, are less viable because the sites below the dam have longer flood durations, although of lower peak flows.

Edge effects

The river itself defines two edges of the riparian landscape element. Cutbank edges and point bar edges are the two distinctive edges of meandering rivers. In riparian forests, both of these edges share features with the forest edges that have been an important aspect of the development of landscape ecology. Many forest edges which have been studied were created by human activity, but natural edges, such as those created by fire or sharp gradients in moisture, or river edges, share common properties with them (Ranney 1977). The most notable factor that has been cited as creating the conditions of edges is microclimate and the effects of increased solar radiation and wind velocities (e.g. Williams-

Figure 4.7 (*a*) Interior of a riparian forest contrasted with (*b*) a cutbank edge site; Cedar River, Iowa.

Linera 1990). The age and maintenance of the edge is also a factor affecting its physiognomic structure and composition (Ranney *et al.* 1981). Along a river, the edges of a riparian forest have very particular characteristics that may distinguish them from upland edges. Kupfer and Malanson (1992*a*) found that many of the features of cutbank edges in Iowa were similar to those reported for agricultural forest edges in the Midwest, and we found little distinction between the cutbank edge and the edge of an agricultural field on the floodplain. The edge sites had higher tree densities, higher sapling densities, and greater species richness than did a riparian forest interior (Figure 4.7). The cutbank sites, however, did not have a greater total basal area, attributed in other studies to increased productivity, than did the interior sites; interior sites were similar to interiors of uplands, but the edge sites had only half the basal area reported for Midwestern upland edges (Ranney *et al.* 1981). Kupfer and Malanson (1992*a*) hypothesized that this result was due to the active erosional process of the river. As the cutbank migrates into the forest, the additional light allows the establishment of many additional seedlings and saplings, which do not significantly increase the basal area. Before additional growth does increase the basal area, renewed channel movement removes individuals and the edge itself migrates toward the

(b)

interior. Kupfer and Malanson (1992*a*) also found greater variability in stem density along the riparian edge than along the agricultural edge, also attributed to the process of cutbank erosion. The cutbank edge sites were probably not in equal stages of maintenance by the river, as evidenced by the curvature of the cutbank and the presence of trees recently toppled into the river. The heterogeneity of the cutbank edge habitat may also play a role because the river is cutting into areas that contained different aspects of the effects of channel dynamics discussed above, e.g. ridges and swales.

The processes by which riparian edges along cutbanks develop seems to be retrogression. Kupfer and Malanson (1992*b*) used a computer simulation model of forest dynamics to investigate retrogression at the riparian forest edge. We altered the light, soil, and water table parameters in the model through time as would be indicated by the retrogressive processes hypothesized to operate when a river channel slowly migrates into a riparian forest at a cutbank. The simulation produced a forest similar to the riparian forest interior before these parameters were changed. Then following change the simulation projected successive forest composition that at first shared compositional characteristics with both the interior and the edge, and this was followed by another stage when the projections became very similar to the edge sites, where they stabilized (Figure 4.8). The simulation indicated that as

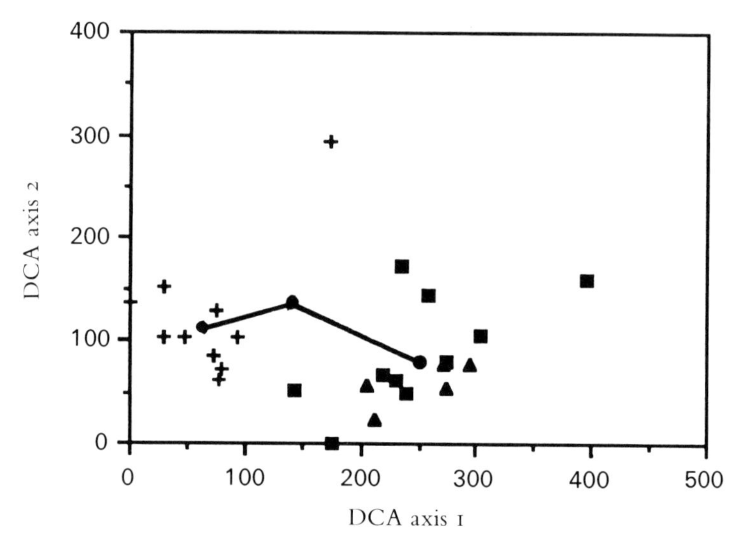

Figure 4.8 The track through ordination space of a simulated interior riparian forest stand when the physical conditions in the simulation are changed to those of an edge site. The simulated stand moves from the center of the ordination locations of observed interior stands to the locations of observed edge stands.

the cutbank approaches edge species become established with the presence of additional light, and this middle stage is maintained while the mature interior trees are still alive. As the conditions continue to change, the interior trees begin to die, and the abundance of edge species increases. Because the simulation model projects a change in community composition from that of an interior riparian forest to that of a riparian edge when retrogressive forcing functions, i.e. the opposite of assumed successional changes, are used, we concluded that retrogression was the best explanation for the temporal process at these edges.

Point bars present completely different situations in which the processes of establishment and succession are central, but studies of this process have relied on analogies to successional processes in old fields, and not to the spatial character of the point bars as edges.

Moisture conditions

Moisture conditions should be important because riparian areas include sites that range from extremely dry to saturated conditions (usually seasonal extremes). On extremely dry sites plant species must be able to take advantage of seasonal wetness, especially for establishment, and must survive the dry season. On saturated sites, the limiting factor is anoxia, and only species that can tolerate anoxic conditions, at least temporarily, can survive. This range of conditions provides an environmental gradient on which species can separate. Differences in moisture conditions, being primarily affected by topography – in turn affected by the channel dynamics discussed above – lead to complicated spatial patterns of vegetation–environment relations on active floodplains (Buchholz 1981). Sedgwick and Knopf (1989) described a condition in which the growth of *Populus sargentii* led to deepened channels and higher terraces, which mean a deeper water table and thus lower rates of seedling survival such that the species will not replace itself on the same site.

Two related physical conditions are relevant and operate differently: actual soil moisture and depth to the water table. These two conditions are related because a high water table can produce high soil moisture and vice versa. Low water tables levels may exist, however, when soil moisture is at least seasonally high. These two factors act together in the former case: when both are high they both might directly affect the uptake of oxygen by roots, and their separate effects would be difficult to determine. When soils are saturated, the water table is effectively at the surface. The water table may act independently of soil moisture when it is at greater depth and can be tapped by a few species with deeper roots. In some studies only one or the other of these two factors has been specified, however, even when both may be playing a role. Some of these studies have also included the direct effects of floods, and floods might be considered as part of a continuum of soil moisture and water table levels, but most ecological studies do not include all the complexity (but contrast Wassen *et al.* 1990). I will discuss some aspects of flooding most related to soil moisture and anoxia, particularly in experiments, here, but will address actual flooding in general in the next section.

Soil moisture

Although drought resistance is important in some regions (e.g. Parikh 1989), most studies related to soil moisture focus on more mesic conditions. In describing plant communities along the Des Moines and Missouri rivers Aikman and Gilly (1948) found that soil moisture was the most notable of the edaphic and climatic variables considered in relation to the floristic patterns. In the same area Weaver (1960) found riparian forest types to be most closely related to aeration and water supply in the soil. Differences in species between first-bottom (lowest) and second-bottom (higher) topographic position were attributed to soil moisture, although he did mention channel dynamics in the context of succession. In Illinois, Crites and Ebinger (1969) attributed poor reproduction by *Populus deltoides* and *Salix nigra* and invasion by *Celtis occidentalis*, *Ulmus americana*, *U. rubra*, and *Fraxinus americana* to drying of the area. In North Dakota, Wikum and Wali (1974) found that available water capacity of the soil and other variables related to moisture stress were predictive for tree basal area and herbaceous cover. Wheeler and Kapp (1978) used a detailed topographic map (contour interval 1 ft or 30.5 cm) to relate species distributions to topography, which they inferred was a gradient of soil moisture. Jackson and Lindauer (1978) found a gradient of plant communities from *Salix* spp., to *Salix* spp. and *Populus sargentii*, to open and closed pure *P. sargentii*, to open treeless stands associated with soil moisture. More recently, Abrams (1985) found variation in growth rates among riparian oaks in Kansas due to edaphic conditions. These studies typify the work in this area in that soil moisture is singled out as a key variable although it is probably correlated with several other important variables, and topographic position is emphasized. Adams and Anderson (1980) described a much longer soil moisture gradient from floodplain, to terrace, to uplands.

Studies emphasizing soil moisture are not confined to the central plains of the USA, although this region has been important (e.g. Petranka and Holland 1980; Collins *et al.* 1981; Bush and Van Auken 1984; Van Auken and Bush 1985; Abrams 1985, 1986). Other studies in Alabama (Gemborys and Hodgkins 1971), New York (Huennecke 1982), and Quebec (Tessier *et al.* 1981) also emphasize the importance of soil moisture in controlling the internal structure of the riparian forest. Gemborys and Hodgkins (1971) described species distributions along a single ordination gradient which was related to depth to water table, slope, and pH. They constructed a moisture regime index to coincide with the sequence of leading dominants in the ordination. In Louisiana Tanner (1986) reported distinct associations of species at both the wettest (flooded swamp) and driest (second-bottom terrace) ends of a gradient of soil moisture, and a continuum of distributions for 15 species across the middle part of the environmental gradient on the first-bottom. He identified six topographic types within the first-bottom. In Czechoslovakia, Vasicek (1991), in association with hydrological studies by Prax (1991*a*,*b*), identified changes in herb species composition that followed reduced soil moisture and an increase in humus when river management structures were built, but reported that the most important species did not differ in their distribution. Hermy and Stieperaere (1981) ordinated species in riparian forests in Belgium and found them most related to soil moisture.

A number of studies have investigated the effects of saturated soil on the establishment and growth of riparian trees using greenhouse, growth-chamber, or other experimental procedures; early studies in the eastern USA share chosen species. McDermott (1954) examined growth in saturated soil of seedlings of six species found in riparian landscapes. He found that saturation stunted the growth of the species similarly, but their recovery following drainage differed. Hosner and Leaf (1962) used comparisons of seedlings grown in saturated conditions to those grown in well watered conditions to classify 14 species of the eastern USA into three classes of tolerance to soil saturation. Hosner and Boyce (1962) compared seedlings growing in saturated soil for different lengths of time to distinguish tolerance and to identify possible mechanisms of tolerance. Hosner *et al.* (1965) examined nutrient uptake in relation to soil moisture conditions, with only one (*Nyssa aquatica*) of four riparian species using more nutrients in saturated conditions. Dickson *et al.* (1965), in a similar experiment, found that two of the four (*N. aquatica* and *Fraxinus pennsylvanica*) grew best in saturated conditions. Hook and Brown (1973) found a gradient of root adaptations among species that coincides with their flood tolerance. More recently, Pallardy and Kozlowski (1981) used more sophisticated measurement techniques to relate leaf water potential to soil moisture. Hook and Scholtens (1978) reviewed the tolerance of trees to flooding and, assuming that root adaptations are the most important morphological and physiological responses to flooding, found support for the hypothesis that discrete adaptations and refinements in root structure and function would be correlated with flood tolerance.

In studies that more directly address flooding, but in fact are most closely related to these experimental studies of soil moisture, experiments have been made in which greenhouse and growth chamber pots, usually of seedlings, have been inundated so that surface water completely or incompletely covers the plant. The earliest of these trials was that of Demaree (1932); she examined the viability of seeds and the survivability of seedlings of *Taxodium* submersed for various periods. Hosner (1957, 1958, 1960) examined seed germination and seedling growth of riparian species inundated for 2–32 days and found generally good germination for all, but considerable variation in health and survivorship. McAlpine (1961) flooded *Liriodendron tulipifera* seedlings in large drums and found no effects during the dormant season, but detrimental effects, most noticeable within days of removal from the drums, during the early growing season. DuBarry (1963) reported distinct differences in the germination success of nine riparian species. Loucks and Keen (1973) inundated seedlings of riparian species, in specially constructed ponds, also describing three tolerance groups. Sipp and Bell (1974) compared net photosynthesis of *Acer saccharinum* seedlings completely inundated with those growing in saturated soil; although photosynthesis decreased in both groups over the 10 days of the trial, the complete inundation was more detrimental.

After a period with less interest, this type of study has been renewed with more emphasis on morphological and physiological details. Pereira and Kozlowski (1977) found a sequence of physiological responses to flooding among *Fraxinus pennsylvanica*, *Populus deltoides*, *Salix nigra*, *Ulmus americana*, *Quercus rubra*, and two *Eucalyptus* species. Sena Gomes and Kozlowski (1980*a*,*b*,*c*) examined several species and specifically considered growth responses. Other studies by Newsome *et al.* (1982) for *Ulmus americana*, Tang and Kozlowski (1982) for *Quercus macrocarpa*, and Peterson and Bazzaz (1984) for *Acer saccharinum* follow a similar line. Tsukahara and Kozlowski (1985), by cutting developing adventitious roots of *Platanus occidentalis* seedlings growing in 2 cm of surface water, provided further evidence for this specific adaptation for riparian species. A review by Kozlowski (1984*a*) emphasizes this interest.

Harrington (1987) described the physiological responses of *Alnus rubra* and *Populus trichocarpa* (both riparian species of western North America) seedlings to 20 days of artificial flooding, which included changes in root, stem, and leaf development, but did not find stomatal closure. Pezeshki and Chambers (1986), however, found stomatal closure and reduced photosynthesis for southeastern USA riparian species after the same period of flooding. Day (1987) found differences in growth, root:shoot ratio, and nutrient uptake between unflooded and intermittently flooded vs. continuously flooded seedlings of *Acer rubrum*. Donovan and McLeod (1984) and Donovan *et al.* (1988) added water temperature as a variable (for thermal output from power plants) along with drained, saturated, and flooded conditions for four southeastern USA riparian species. When flooded conditions were combined with temperatures 10 °C above ambient, decreases in growth, biomass and survivorship were observed, although flooding alone or with a 5 °C treatment were not different from other treatments. Jones *et al.* (1989)

combined treatments of inundation and saturation with root competition from established forest vegetation. Inundation resulted in high mortality compared with soil saturation. Hallgren (1989) studied *Populus* hybrids in relation to flooding and found increased development of morphological traits that increase oxygen transport. Hardiyanto (1989) examined clones of *P. deltoides* in relation to water shortage and nitrogen fertilization. Water stress was the most significant variable. Firth and Hooker (1989) found evidence for what might be termed retrogression in floodplain wetland stands subjected to flooding and heated effluents. Undisturbed stands were dominated by *Acer*, *Fraxinus*, and *Nyssa* (species unnamed), while disturbed stands were dominated by *Salix* and *Alnus* species, and had a greater importance of shrubs and herbs. During the year of the study (1986–7) the disturbed sites experienced considerable mortality. Melick (1990*a*,*b*) tested both the drought and flood tolerance of two riparian species from temperate rainforests in Victoria, Australia. Differences in their tolerances may differentiate the niche requirements and distributions of these species.

A number of reviews summarize the effects of soil saturation and flooding on vegetation. Gill (1970) provided a summary of orders of tolerance for species as provided by 12 studies, discussed mechanisms of tolerance, and tabulated the success of different species in riverbank and floodplain plantings, mostly from central Europe. Hook (1984) and Kozlowski (1984*a*,*b*) provided detailed reviews on the morphological and physiological adaptations of plants to soil saturation and flooding. For woody species, Kozlowski (1984*b*) reported on distributions in the USA and types of responses related to flooding, and on genetic, demographic, and hydrological factors. He reviewed studies on regeneration, several aspects of growth, and mortality. In a more detailed examination of responses, Hook (1984) reviewed work on seed germination and dormancy, morphological responses of stems and roots, anatomical variations favoring gas exchange, and anaerobic metabolism and alcohol dehydrogenase activity. In some areas, however, the question is one of moisture shortage.

It is interesting to note that two major groups of studies have been led by individuals, John Hosner at Southern Illinois University and T.T. Kozlowski at the University of Wisconsin. In the late 1950s and the 1960s, Hosner was responsible for a significant body of experimental work which documented the response of seeds and seedlings to soil saturation and flooding. In the 1980s Kozlowski, already well established in the field of water stress, advanced the earlier work, sometimes using the same species, by looking at more precise indicators of functional morphological and physiological responses to anoxia. Conversely, the studies of actual responses of vegetation to flooding at field sites are widely scattered, and while representing some centers (the University of Illinois) are not so dominated by individuals. Now, a Center for Streamside Studies has been founded at the University of Washington. Its first director is R. J. Naiman, who brings to it a landscape perspective.

Depth to the water table

Depth to the water table may in and of itself be an important factor in the internal structure of the riparian vegetation. It can be imagined that saturated soil is in effect a water table at the surface, and flooding a water table above the surface. In some comparative cases the soil moisture may be very closely related to the depth to the water table, and in many cases the variations in soil moisture in one place will be related to this measure. Tessier *et al.* (1981) reported distinctions within a vegetation type, on a riparian gradient of flood duration, related to depth to the water table. In the Netherlands, Boedeltje and Bakker (1980) and Grootjans *et al.* (1988) reported the importance of groundwater regimes to riparian vegetation, and Beltman and Grootjans (1986) specifically referred to landscape ecology in describing the effect of groundwater availability on the transverse pattern of vegetation in riparian areas. Bagi (1987) attributed some of the variation in vegetation he mapped in Hungary to groundwater levels in riparian areas. Popov (1985) documented a gradient of riparian vegetation related to the level of the water table during the growing season, and Schnitzler *et al.* (1991) found Spring levels to be important. Klimo and Prax (1985) diagrammed the relation between precipitation and groundwater (which they graphically linked to above ground levels during flood periods) for 27 months in 1969–71 in a riparian forest in Czechoslovakia. They also noted, however, that perched water tables could be distinct from the general water table, which is more directly connected to the level of the river. Jemison (1989) found that soil moisture could not be used as an indicator of a riparian environment in the absence of riparian vegetation and that the distribution of present riparian vegetation was determined by the water table. Where seasonal variation is important the steadier water table level will be important in association with specific species.

Mitsch *et al.* (1979*a*) studied the relation between river flow and groundwater in a riparian forest area in Illinois. Their aim was to quantify some relations between floodplain hydrology, water quality, and sedimentation. Their measurements of the level of the water table were not sufficient to model the relation to flooding, river stage, and precipitation. Also, their chosen floodplain seems topographically simple, and although they mention that some topographic variation exists they did not include it in their study design, nor did they quantify the soil textures that might be associated with these topographic differences. Any attempt to relate soil moisture or the storage of water in floodplain soils without this information is likely to be misleading. The challenge of an adequate sample design when considering all of these effects has limited my work on this topic in Iowa.

In an unusual study, Dawson and Ehleringer (1991) used oxygen isotope ratios to investigate the source of water for streamside trees. They found that smaller trees (less than 20 cm DBH) used soil moisture as a source while larger trees used deeper water sources, presumably groundwater. This study is interesting because it belies an unusual conception of basic hydrology. The authors stated that it was commonly assumed that riparian trees used water from the stream as their source. This contention is untrue. In most riparian areas the direction of flow is from the soil water and groundwater into the stream (this process is the basis for many studies in the filter effects of riparian forests discussed in Chapter 5). The variable source area concept amplifies this relation (Hewlett and Hibbert 1967). Water from the stream should move into the soil only under influent conditions, which are usually only a characteristic of exotic streams in arid areas or temporarily in other areas when a flood pulse raises the level of the river above the hydraulic base level. Dawson and Ehleringer (1991) did not recognize this relation, and they did not provide any information about the hydrological and topographical system of their study site, so it is not possible to determine the likely significance of their results. They assumed that because the water in the upper soil layers (10–50 cm) was similar in isotope ratio to that in the stream, water had moved from the stream into the soil. Whether or not the water had moved into the surface soil from the stream in this case, which is possible, the conclusion that larger trees in riparian areas use deeper water sources, probably groundwater sources, should come as no surprise to anyone.

Studies in riparian groundwater hydrology have attempted to relate water table levels to precipitation and/or river levels (e.g. Bell and Johnson 1974; Granneman and Sharp 1979; Doty *et al.* 1984; Sophocleous *et al.* 1988). Kondolf *et al.* (1987) studied hydrological responses to

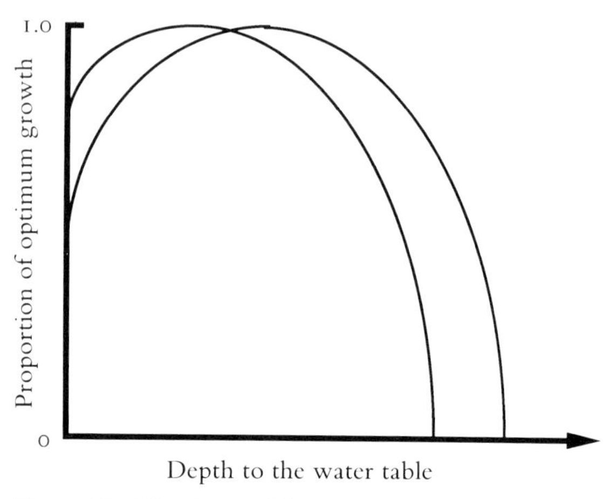

Figure 4.9 The form of the response function for tree growth relative to water table depth used in the simulation models developed by Phipps (1979) and Pearlstine *et al.* (1985).

flow diversions in relation to riparian vegetation. They distinguished relations between geomorphic features that controlled the hydrology and the pattern of vegetation. Space (1989) documented different sources of groundwater input to riparian vegetation depending on the geology and geomorphology of the basin. Girault (1990) specifically illustrated the use of piezometers to characterize forest communities in areas influenced by the water. In a similar effort in Iowa I have not been able to easily relate water table levels to river stage and precipitation because successive years of extreme drought and extreme flooding disabled the piezometers, but the measures made are the basis for some of the concepts expressed in Figure 4.1.

Such studies of the level of the water table in relation to riparian vegetation are important. In simulation models Phipps (1979) and Pearlstine *et al.* (1985) have used equations in which the growth of riparian tree species is a function of depth to the water table. Pearlstine *et al.* (1985) cited Phipps (1979) as the source of their equation, but in fact use coefficients that are quite different. In both cases an optimum depth to water table is defined for each species (Figure 4.9). Optimum growth is reduced as conditions vary from this optimum. Pearlstine *et al.* (1985) included much greater sensitivity to suboptimal conditions in their model than did Phipps (1979), and this may explain the drastic change in species that they project as a result of an engineering diversion. I have adapted this model for use in Iowa and am now examining differences in these coefficients more closely. Using Phipps' (1979) coefficients, only extreme changes in flood regime have major effects on species dynamics (Liu and Malanson 1991). In reality, the shape of the curve as well as its modal point should vary for each species.

Flooding

Given that the lateral flow of water is a distinguishing feature of riparian areas, flooding has been examined as a major causal process. This process, however, is often not distinguished in its actual operation. In addition to the disturbance effects related to channel dynamics and discussed above, flooding can affect plants either through direct mechanical effects or through oxygen depletion; for example, Stevens and Waring (1985) reported burial and drowning for a case study on the Colorado River. The latter effect thus relates riparian processes to the more general class of wetland responses to anoxic conditions. The direct mechanical effects of flooding are not well documented, and may in part also be related to channel dynamics on the floodplain. Many studies relate the internal structure of riparian forests to the frequency, duration, and season of flooding without distinguishing between mechanical and physiological effects. As noted above, flooding is directly related, as an input, to soil moisture and depth to groundwater (e.g. Nortcliff and Thornes 1988), but soil type may be less important (Bowman and McDonough 1991). Because these factors are more difficult to document than the frequency and duration of floods, which can be determined from river gauge records, flooding has been used, ether explicitly or implicitly, as a surrogate (e.g. Chambless and Nixon 1975; Metzler and Damman 1985). Where possible I will infer which effect was the object of study, but in some cases the object was not so specific.

A number of studies attempting to summarize information about the effects of flooding on riparian vegetation and on the flood tolerance of individual species have been produced by agencies of the US government. These are often related to river regulation projects. US Army Corps of Engineers researchers have been especially active in the southern USA. Whitlow and Harris (1979), Huffman (1980) and Huffman and Forsythe (1981) analyzed the flood tolerance of many species. Klimas *et al.* (1981) summarized some of this information and presented an extensive review of effects on wildlife, briefly discussed below. The US Fish and Wildlife Service has also produced a series of documents on effects of flooding. Teskey and Hinckley (1977*a*) briefly summarized the responses of vegetation and soils. Subsequent volumes focused on vegetation regions of the USA (Teskey and Hinckley 1978*a*,*b*, 1979*a*,*b*,*c*; Walters *et al.* 1980; Lee and Hinckley 1982), and Chapman *et al.* (1982) provided an index and addendum. While many of these volumes focused on identifying a tolerance level for individual species and a successional sequence for ecoregions, the latter also extended the tolerance information to give some idea of a range of tolerance for many species, and they discussed planning, management, and restoration aspects of riparian systems ecology.

In addition, in some forests deep flooding can totally submerge saplings and is a direct factor, not necessarily an extension of soil moisture. Campbell *et al.* (1992) reported habitat distinction among species related to level of inundation in Amazonian varzea, where flooding reached depths up to 4 m.

Mechanical injury and/or site fertility

One of the effects associated with channel dynamics, though not necessarily with actual changes in the channel, is the mechanical injury to plants during flooding. In mechanical injury, individuals may be felled, scarred, defoliated (saplings), washed away (seedlings), or buried. The deposition of sediment may also alter the fertility conditions of a site. While fine sediments, especially those eroded from agricultural fields, may be high in nutrients (see Chapter 5), coarse deposits may lead to nutrient poor and dry conditions.

Different abilities to withstand this stress or to recover from damage allow species to spatially segregate among areas where the action of flood differs (e.g. Johnson and Bell 1976*a*). Miller and Johnson (1977) cited the ability of *Alnus rhombifolia* to tolerate mud and rock flows coming into the riparian zone from steep surrounding slopes. Nilsson (1987) considered the variation in species diversity in aquatic and riparian sections of transects in relation to the midstream velocity of a stream. He found that while the aquatic community was most diverse at intermediate levels, the riparian community was most diverse where current was strongest; he noted that this measure of current may not be the best indicator of flood damage. Roberts and Ludwig (1991) distinguished four riparian vegetation types of floodplain wetlands in Australia, the species compositions of which were correlated with flow velocities and exposure to wave action. Both of these variables, while they imply direct injury to plants, may also operate by affecting the fertility of a site, as shown for lakeshores by Keddy (1984).

Gawler (1988) found both the disturbances of flooding and ice-damage and the stress of soil moisture shortage to affect the fecundity and growth of *Pedicularis furbishiae* (Furbish's lousewort) along the St. John's River, Maine. Ellery *et al.* (1990) also considered current velocity and the direct effects of flooding on channel vegetation in the Okavango Delta. The direct effects of flooding have been explored, especially in respect to the regeneration and growth of trees. Broadfoot and Williston (1973) described adverse effects of sediment deposition and, for seedlings, of complete submersion. Kennedy and Krinard (1974) reported complete destruction of some seedlings and failure to sprout of some planted cuttings that had been inundated by 8 m of water by a flood on the Mississippi River, but they found that siltation of 1.5 m depth had not harmed plants over 2 yr old. Different responses along flood gradients have been observed (e.g. Sipp and Bell 1974; Johnson and Bell 1976*b*; Sharitz and Lee 1985). In Iowa, we observed no difference in the numbers of surviving individuals of several species between those established in years with or without flooding, and inferred that differential flood effects did not exist (Zipp 1988). While the coarse differentiation found using dendrochronological methods may have been a problem, it seems that seedlings of most species established in nearly every year in an area that experienced a flood return interval of two years. Nilsson and Grelsson (1990) also found that the redistribution of litter by flooding also affected species composition. More complex interactions have received little direct attention, however.

Day *et al.* (1988) presented a model relating flooding, including the effects of ice and the importance of litter removal on fertility, to riparian marsh vegetation. The type of model presented by Day *et al.* (1988), with the appropriate observations, is needed for the wider range of riparian vegetation and the more numerous environmental factors that may have some effect.

Atkin (1980) emphasized the establishment phase. He reported distinct differences in the age structures of riparian sites in New Jersey, which led him to conclude that establishment was primarily a post-flood phenomenon. This was most true for his advancing point bar site. He also studied the establishment of trees either by seedlings or vegetatively, and found that few seedlings survived to the sapling stage. Most saplings had originated as vegetative sprouts. He concluded that seedling survival would be difficult because of the mechanical stress of flooding.

Bush and Van Auken (1984) described the internal structure of riparian forest in Texas, and reported a general ecocline away from the river related to elevation and flood frequency. They found that the diversity increased away from the river where upland species became associated, and that tree-falls caused by flooding were important in determining density and composition of the forest near the stream. Hukusima *et al.* (1986) reported the maintenance of a mosaic of vegetation due to the direct damage of flood water, indirect damage and opportunities presented by alluvial deposition, and the effects of erosional channels.

Studies by Hupp (1982, 1983; Hupp and Osterkamp 1985) explicitly emphasized the relation between processes of fluvial geomorphology and patterns of vegetation in Virginia. The general concepts were not greatly different from those of other researchers, but the more detailed examination of processes surpassed the simple idea that vegetation at different elevations is differently affected by flooding. The concept that the floods create the differences in elevation was reiterated clearly. Hupp identified four major fluvial landforms in this region: depositional bar, channel shelf, floodplain, and terrace. Illustrations of these show a hypothetical distinct transect from the channel to the surrounding hillslopes. He presented a hypothetical gradient of vegetation associated with the lowest channel features (Hupp 1983), and more quantitatively Hupp and Osterkamp (1985) contended that the association between species and locations was due to flood frequency and duration rather than to associated sediment characteristics. Osterkamp and Hupp (1984) added larger rivers and argued that stream power controlled these geomorphic surfaces and that species were specifically associated with them. Hupp (1988) reviewed these and other studies in relation to paleohydrology. While they are somewhat repetitive, they indicate that some aspect of flooding, not operating through its effects on soil moisture or sediment composition, may play a role in the distribution of riparian plant communities within a landscape element.

Oxygen depletion

Although this mechanism could be considered as an extension of a gradient of soil moisture, many studies simply rely on a measure of above-ground flooding as a variable. Saturated soil conditions in the pre- or post-flood period are usually ignored in these studies. With respect to oxygen depletion, early studies include those of Shunk (1939), Yeager (1949), Parker (1950) and McDermott (1954) who considered the effects of saturation on riparian forests, and the more general report by Kramer

(1951). Chavan and Sabnis (1960) considered seasonal changes in ephemeral vegetation in comparison with perennial vegetation, which is less affected by regular flooding. Duvigneaud and Denaeyer-De Smet (1970) diagrammed a cross section of vegetation and described its relation to river level during the growing season. McKnight *et al.* (1981) cataloged flood tolerances of species in the southern USA and described related traits. More recently, Blom *et al.* (1990) described adaptations of two riparian genera in the Netherlands to flooding. In almost all cases, flooding differentiates among species because of their degree of tolerance to poor growing conditions; conversely, Rikard (1988) found that for *Nyssa aquatica* and *Taxodium distichum* productivity increased when flooded more frequently.

The reason for the general approach of the studies cited above is that most such studies actually relate vegetation to floodplain topography. The frequency and duration of flooding is inferred by elevation to topographic zones such as first- and second-bottoms or terraces. It was in this way that Buell and Wistendahl (1955), Wistendahl (1958) and, later, Frye and Quinn (1979) examined riparian vegetation of the Raritan River, NJ, distinguishing sand bars, levees, inner floodplain, and terrace features. In addition to identifying major topographic features, Buell and Wistendahl (1955), specifically discussed the importance of small topographic forms of ridges and sloughs in what they termed the inner floodplain, farthest from the river and less well-drained. They noted that these features create different conditions for establishment and thus influence the areal differentiation of the vegetation. They also noted the importance of the depositional environment in determining soil moisture conditions. Frye and Quinn (1979) documented the edaphic factors which seemed to be controlling the distribution of tree species most directly, and Buchholz (1981), following up on the ideas of Buell and Wistendahl (1955) examined the minor drainage pattern of ridges and sloughs and its effects. In Alaska, Woodward-Clyde Consultants (1980) studied the effects of mining riparian gravels, and found that small topographic variations in the mined areas affected the vegetative recovery and diversity. Huennecke and Sharitz (1990) found that substrate heterogeneity affected the regeneration of *Nyssa aquatica*, and in Florida, Titus (1990) has documented the spatial differentiation of seedling establishment within a floodplain in relation to microtopography, including the conditions created by tree-falls. Malecki *et al.* (1983) also commented on the importance of the microtopography of tree-falls.

Hosner and Minckler (1963), in a study related to experimental work, documented seedling abundance in riparian areas in Illinois. They identified the environmental conditions and associated species and canopy conditions in which seedlings thrived. Bedinger (1971, 1978, 1979) emphasized this topographic connection in studies in Arkansas. Although these researchers noted that groundwater relations, soils, and drainage were also important hydrological factors, he concluded that topographically controlled flood frequency and duration was the primary factor influencing the internal structure of riparian forests. He later noted that more work was needed to relate river stage to the processes of flood stress (Bedinger 1981). Kennedy and Krinard (1974) reported that standing water had killed some species following a flood in Mississippi. Malecki *et al.* (1983) found that in cases of controlled flooding the additional differences caused by different soil drainage properties were noticeable. Lee *et al.* (1985) found that a distinct moisture gradient affected riparian species in montane areas of Montana, but noted that it was only one of several important factors. Tanner (1986) differentiated the distributions of species in Louisiana. He found distinct groups of species on terraces and almost permanently inundated swamps, but a continuum of species on a gradient of soil moisture on the primary floodplain. Bell (1974*a*,*b*, 1980; Bell and del Moral 1977) reported spatial differentiation related to flooding in the tree, shrub, and herb strata of riparian forests in Illinois. These, too, are related to the frequency of river stages. Also in Illinois, Robertson *et al.* (1978) related the distributions of species to flood duration and depth, but they also considered soil moisture.

While almost all of the studies reported above concentrated on woody vegetation, interesting observations on herb layers within floodplain forests have also been reported from Wisconsin. Barnes (1978) found that the distribution of herbaceous species was correlated with elevation above the river, which he inferred indicated the effect of the frequency and magnitude of flooding. Menges and Waller (1983) used the model of plant strategies proposed by Grime (1979) to analyze the distribution of herbaceous species in relation to gradients of light and elevation, with elevation again being an indicator of flooding. They noted that flooding would have the effect of a disturbance on short-lived species and that of a stress on long-lived species. They found that light was only secondarily important in this environment. Later, Menges (1986) reported an ordination analysis of floodplain herbs and a study of diversity that was correlated with elevation. Shelyag-Sosonko *et al.* (1987) reported a very complex meadow vegetation subject to inundation on floodplains in the Ukraine.

In California, Hanes *et al.* (1989) described three developmental phases of riparian vegetation in Mediterranean southern California which they termed alluvial scrub vegetation. Their description implies a successional sequence related to flooding as a disturbance. Bren (1987, 1988; Bren and Gibbs 1986; Bren *et al.* 1988) made a detailed analysis of inundation and its effect on a *Eucalyptus amaldulensis* forest in southeastern Australia. Although his primary purpose was to estimate flooding based on vegetation, the relations described are appropriate to studies of riparian ecology. Dunham (1989*a*) ordinated and classified riparian species distributions of the Zambezi River. He found a developmental sequence related to disturbance and differentiation based on soil texture and flood frequency. Furness and Breen (1980) described six communities and two subcommunities related to periods of inundation in South Africa.

Some of the effects of riverine processes are also seen in tropical areas. While Bacon (1990), Bruenig (1990), Myers (1990), and Paijmans (1990) were focused on wetland forests, some of these include riparian forests, as revealed by their discussion and illustration. In their photographs and diagrams, they reveal that river levees and river flooding can both be important aspects of wetland forests in these areas. In some areas of extensive basin and coastal wetlands, the higher levees present important areas for the distinction of the internal structure of these habitats. Frangi and Lugo (1985) documented ecosystem cycles and flows in detail for a small floodplain area in Puerto Rico, concentrating on productivity and the export of matter to the stream.

Brown and Peterson (1983) reported differences in species composition and diversity, biomass, tree growth, and litterfall between riparian sites that were inundated during spring floods and one inundated most of the time. The site of short-duration floods had lower values in every category. This forest site was less productive than similar sites elsewhere, indicating that not all such sites are very productive. This study also indicates that conditions of extreme anoxia are not always limiting in comparison to better aerated sites. In a contrasting study, Rice (1965) found vegetation correlated with precipitation in Oklahoma, and discounted the effects of flooding, which he considered to be too limited in duration. Collins *et al.* (1981) updated this study with the same conclusion. In Iowa we have observed much better correlations of tree-ring width of *Salix nigra* with both precipitation (positive) and temperature (negative) than with duration of flooding. Lee (1945) also considered macroclimate to be more important than local factors in controlling the composition of a riparian forest in Ohio, but did not really examine soil moisture.

Hydraulic structures

In Chapter 2 I noted that dams, canals, and dikes were factors leading to the destruction of riparian vegetation. Studies emphasizing a deliberate change in the flood regime due to dams, diversion, or channelization also note the importance of channel dynamics, mechanical effects, and moisture availability or anoxia. Some studies which I discussed above in the context of channel dynamics are particularly relevant again here, but I will focus on a few others that are especially relevant for hydraulic structures.

Dams

In-depth studies of the effects of a dam on downstream riparian vegetation have been conducted by Johnson and associates (Johnson *et al.* 1976, 1982; Johnson and Brophy 1982; Keammerer *et al.* 1975; Reily and Johnson 1982). Johnson *et al.* (1976) identified declines in the establishment of *Populus deltoides*, which they attributed to a lack of channel dynamics which would create the needed seed bed, and in the establishment and growth of *Acer negundo*, *Fraxinus pennsylvanica*, and *Ulmus americana*, which they attributed to a decrease in flooding which formerly brought needed moisture to the terraces sites of these species. Reily and Johnson (1982) extended the study of tree growth and found sensitive responses to the change in hydrological regime. For example, the growth of *P. deltoides* had been correlated with river flows before the dam was completed, and has since been correlated with local precipitation. Johnson and Brophy (1982) summarized the effects of the dam. They noted that growth declines were dependent on the particular species and its location on the floodplain, and that reduced channel dynamics and bar formation has reduced the general importance of *Populus* and *Salix* communities which were in essence fugitive species dependent on new sites. Conversely, Turner and Karpiscak (1980) and Williams and Wolman (1984) reported increases in riparian vegetation downstream from dams. Nilsson *et al.* (1991*a*) found a reduction in species diversity related to river regulation in Sweden. These studies illustrate the interaction between flooding, levels of the water table, soil moisture, and channel dynamics, which can be controlled by impoundments.

Keammerer *et al.* (1975), Johnson *et al.* (1976) and Reily and Johnson (1982) found that the reduction of spring flooding had detrimental effects on the riparian vegetation downstream of dams in North and South Dakota. Rood and Heinze-Milne (1989) documented a drastic decline in riparian forests in nearby Alberta. Rood and Mahoney (1990) summarized the key processes of downstream change in these riparian areas. The primary factor was hydrological change of reduced flooding and spring flows, which led to drought stress; the secondary factor was geomorphological change of reduced meandering and sediment deposition, which inhibited the creation of necessary seed bed conditions. They recommended that properly managed flows from reservoirs could reduce their impacts.

It seems that the vegetation along the Platte River in Nebraska, however, has greatly increased in recent times because of upstream impoundments which limit floods (Williams 1978; Lindauer 1983; Crouch 1979*a*; Knopf and Scott 1990); Martin and Johnson (1987) reported a similar change along the Medicine Lodge River in Kansas. Sedgwick and Knopf (1989) reported, however, that on the South Platte the populations of *Populus sargentii* were declining because deepening river channels led to lower water tables relative to the terraces on which seedlings had established, which led in turn to high mortality at this stage. In an another reversal of effects, Szaro and DeBano (1985) found that dams that increased flow in intermittent streams in the southwestern USA increased the abundance of riparian vegetation, but Fenner *et al.* (1985) found that the changes in timing and magnitude of floods had a negative effect on the regeneration of *Populus fremontii*, a riparian dominant in this region. Likewise, Stevens (1989; Stevens and Waring 1985) found that by altering substrate quality the Glen Canyon Dam had reduced riparian tree establishment in the Grand Canyon.

Diversion projects in California, where such projects are prolific (cf. Harris 1986), have been a source of information, and several studies reported in two symposia on riparian vegetation in California examined the effects of flow diversion projects. Stine *et al.* (1984) documented vegetation change following diversion projects in eastern California during this century, reporting almost total loss of riparian vegetation in many areas, and Brothers (1984) documented how such losses of native riparian species led to successful invasions of riparian exotics (see Chapter 6). Langley (1984) described potential losses from even a small diversion project. Several studies specifically examined the ecophysiological responses of species to changes in available water (Smith *et al.* 1989; Nachlinger *et al.* 1989*a*; Leighton and Risser 1989; Dains 1989; Williams 1989). Nachlinger *et al.* (1989*b*) described current vegetation conditions in relation to a diversion project, and Stromberg and Patten (1989) described the recovery of vegetation in an area in which flow has been restored. Stromberg and Patten (1990) later presented a methodology for determining the flow requirements of riparian vegetation that could be applied in cases of diversion. Harris *et al.* (1987), however, have commented that stretches of riparian vegetation respond individualistically because of complex relations between vegetation and the environment in reaches where streams are or might be diverted.

Pearlstine *et al.* (1985), as discussed above, modeled significant impacts from a river diversion over the course of a century. The coefficients that they used for species response to water table levels were, however, more sensitive than those proposed by Phipps (1979) for this type of model. In papers related to another body of work not discussed here, i.e. the effects of inundation at the margins of reservoirs (see also Green 1947), Franz and Bazzaz (1977) and Austin *et al.* (1979) presented models of vegetation change for managed flood regimes as applied to some of the areas studied by Bell (1974*a*,*b*, 1980) and Hanson *et al.* (1990).

Considerable work has been done on the impacts of hydroelectric development on the Zambezi River in Africa (Pinay 1988). Attwell (1970) pointed out that the Kariba dam had led to altered flood regimes, which resulted in vegetation and change and which directly and indirectly stressed the fauna. Changes in the sediment dynamics of the river were reported by Guy (1981). Such impacts are now being recognized and taken into account in planning new developments (Du Toit 1984). Sheppe (1985) reported downstream effects on an interacting system of riparian vegetation and grazing ungulates in Zambia, and Dunham (1989*b*, 1990*a*) reported on the changes in riparian vegetation found downstream of the Kariba dam after closure. In addition to the effects of flooding regime cited by Sheppe (1985), he also noted that changes in the human population density of the area and associated changes in the fire regime, plus secondary changes in the populations of termites and large mammals, had significant effects on the riparian vegetation. This study indicates the complex ecological relations which dams can alter. Conversely, Zinke and Gutzweiler (1990) have proposed means of improving riparian habitats in concert with flood protection dams in Germany.

These studies are directly related to the experimental and field studies of natural systems because the processes affected are those cataloged above: channel dynamics, moisture conditions, and flooding. For example, Smith *et al.* (1991) contrasted the physiological responses of riparian vegetation along paired diverted and free flowing streams in California in these terms, and their results are comparable to the physiological studies mentioned above.

Channelization

In what should have been a national review, but which was confined to cases in a few states of the USA, the Council on Environmental Quality (1973) produced a report on the impact of channel modifications. These modifications were essentially straightening to increase drainage. When a channel is straightened the result is that stream power is concentrated; given a low enough base level, the stream will incise and create a deep channel. This effect has been observed throughout the midwestern USA (e.g. Daniels 1960). The floodplain vegetation is then left on a high terrace that might never be flooded (cf. Graf 1985 for arroyo development in the southwestern USA) (Figures 4.10 and 4.11). Some consequent changes in riparian vegetation have been reported by Fredrickson (1979) in Missouri, and Ellis and Whelan (1979) reported negative consequences for some bird species in Virginia. Conversely, Ulmer (1990) found no adverse effects on the floodplain forest of a section of the Tennessee–Tombigbee canal. He made detailed measurements of soil moisture and water table levels and found that although these factors were affected, the forests, growing on clay soils, were largely insulated from any changes and normal growth rates had continued subsequent to the construction of the canal. In another detailed study in Northern Ireland, however, Wilcock and Essery (1991) found that channelization, while not greatly affecting the water table at a distance, does remove water from storage in gravels near the river, reduces the number of low-flow days, and reduces the input of water and nutrients to riparian wetlands because of fewer over-bank flows.

It is likely in such a case that a succession to species characteristic of drier sites would occur. Whether or not such species were within easy dispersal range would depend on the landscape structure of the riparian system. Along the Rhone and its affluents in France, the meaning of channelization has not always been straightening, but some of the effects are the same. Bravard (1987) documented some of the history of dikes used to keep the river within confined channels. These structures and the

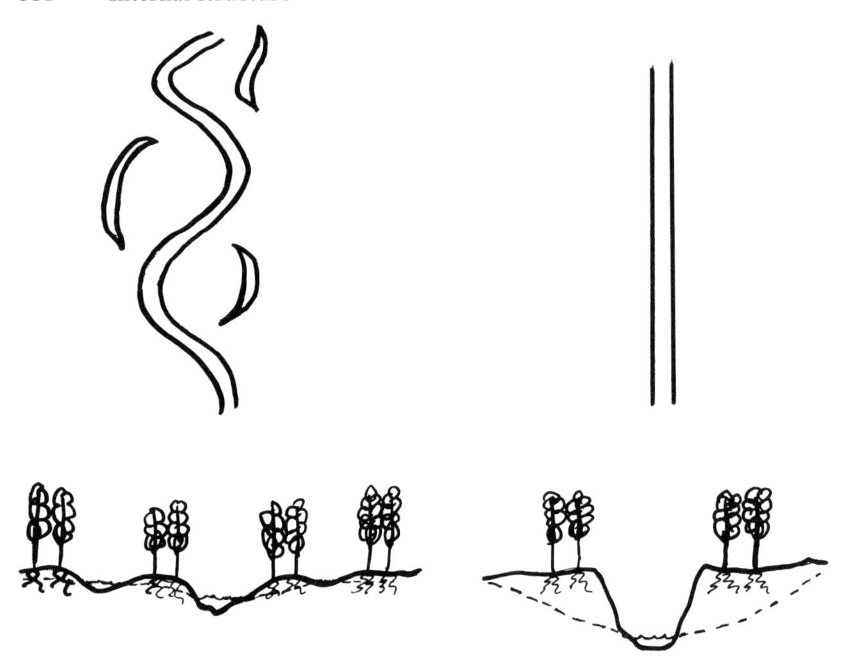

Figure 4.10 When a channel is straightened the plan and cross sections both change, and the relation of the water table to the edge vegetation can be altered.

protection of the bank by rip-rap (see below) or walls, especially in urban areas (e.g. Fortune 1988), have led to an isolation of the river from the floodplain similar to that which I have suggested for agricultural areas in Iowa. G. Pautou (personal communication; cf. Pautou *et al.* 1991, 1992) has observed that what were formerly floodplains in some places in France are now being dominated by distinctly upland tree species. This transformation has taken place over centuries, and immediate change in a place like Iowa may not be dramatic. Without the protection afforded the banks as in France, the straightened channels in agricultural areas eventually begin to modify their form by developing new meanders (Noble and Palmquist 1968). Simon and Hupp (1986, 1987) found that ongoing processes following channelization lead to stabilization; these geomorphic processes interact with vegetational change. In this case complexity within the deeper channel may develop. Hupp and Simon (1991) described a sequence of events for straightened channels in Tennessee which lead to a more natural channel bank environment. After degradation of the channel, slumping leads to widening, and then an aggradational phase begins which includes point bar development

Figure 4.11 A straightened channel has incised into the floodplain c. 5 m, and some bank collapse is evident; White Breast River, Iowa.

and meandering. Deposition provides sites for colonization by pioneer species such as *Salix nigra*, and the combination of depositional surfaces and vegetative growth lead to a new quasi-equilibrium channel form.

In some areas of straightened channels and elsewhere a variety of forms of 'rip-rap' revetments (stone blankets of cobble and larger clasts, or even abandoned autos!) are used to reduce bank erosion. While these alterations initially destroy the natural bank protection afforded by riparian vegetation, they are occasionally subject to colonization (Shields 1991).

Multiple effects

Direct and indirect impacts of people have altered the riparian ecology of the Rhone in France for 7000 years. While deforestation in the drainage basin was the first and indirect effect, later channelization, diversions, gravel mining, and damming have had significant direct effects, and these have accelerated in the twentieth century (Roux *et al.* 1989).

With particular attention to recent hydroelectric developments, which include dams, diversions and embankments, Pautou and Amoros and colleagues (e.g. Amoros *et al.* 1987*a*,*b*; Pautou 1983, 1988; Pautou and Bravard 1982; Pautou *et al.* 1991) have taken on the challenge of determining the factors affecting the ecological dynamics in a complex environment, and have incorporated a landscape ecology approach. One key factor has been the use of extensive large-scale mapping (1:5000) as a basis for studies. They have noted that some impacts may lead to irreversible vegetation dynamics, i.e. to new and different communities from those found in the natural landscape (cf. Decamps *et al.* 1988). Furthermore, Pautou *et al.* (1991, 1992) have addressed the difficulties of predicting the results of environmental impacts in such a setting. They noted that the possibilities of deviation and innovation, whereby new conditions allow species to interact in different ways and create different communities, make realistic prediction difficult if not impossible, but they stress the importance of understanding the long-term dynamics that underlie the short-term responses to impacts.

Considerations of such impacts are not confined to the USA and France; e.g. Walker (1985) reviewed such impacts in Australia, and human impact has been the focus of ecological studies of riparian vegetation in Belgium (Hermy and Stieperaere 1981).

Wildlife ecology

Beaver (*Castor* spp.)

The effect of beaver in the riparian landscape is extensive (Duncan 1984; Naiman *et al.* 1988*c*), and these animals specifically alter landscape relations (Johnston and Naiman 1987). The most notable natural hydraulic structures are beaver dams (Figure 4.12). Beaver alter the flood regime of their environment much as do human-built dams, but their structures are of course smaller and the effects scaled down. Importantly, the effects are temporary. Dams break or food resources change, and beaver move (cf. McIntyre 1981; DeByle 1985). The former beaver pond fills with sediment, and succeeds toward a terrestrial riparian habitat. The downstream effects of the original dam are mitigated. The beaver build a dam elsewhere. Gill (1972*b*) described a cyclic relation between beaver, riparian vegetation, and fluvial geomorphology. The specific type of flow associated with meanders leads to a particular type of sediment deposition on point bars with meander scroll depressions. This environment is favored by *Populus balsamifera*, which is a preferred food of beaver. The use of this species by the beaver leads to succession to *Salix* and *Alnus* species, whereupon the beaver abandon the site. *P. balsamifera* then regenerates on the site and the beaver later return. Beaver more generally affect the diversity of riparian habitats. Allred

Figure 4.12 One of several sturdy beaver dams on a small tributary to Otatso Creek, Montana, has created an extensive wetland.

(1980) documented increases in habitat types by beaver activity in Idaho. Dahm and Sedell (1986) reported that the beaver activities, in part by altering the anaerobic conditions, led to diversity. Nummi (1989) simulated the hydrological role of beaver by building a dam on a small stream in Finland (later, introduced beaver, *C. canadensis*, moved into his study area!) and found signs of stress in riparian trees subjected to inundation. Pastor *et al.* (1991) have reported extensive modifications of riparian landscapes caused by an increase in beaver populations in northern Minnesota during the past 50 yr.

The effects of the removal of beaver on riparian ecosystems have been noted in the USA, where records of beaver hunting and trapping can be used to document the events (Parker *et al.* 1985), and re-occupations are having a reverse effect (Brayton 1984; Apple 1985). Butler (1991*a*) has cited their re-occupation in altering the landscape of the southeastern USA. He gave a specific example of their alteration of the hydrogeomorphology of a stream (Butler 1989), and a review of these processes (Butler 1991*b*).

Beaver also affect the internal structure of riparian landscapes in a way not related to their hydraulic structures. Beaver selectively cut trees (Jenkins 1975, 1979, 1980). This selective harvesting may lead to a difference in the community composition of a riparian forest. Barnes and Dibble (1988) found that beaver greatly reduced tree density, and that food species such as *Fraxinus pennsylvanica*, *Carya cordiformis* and *Celtis occidentalis* would suffer in relation to *Tilia americana* and *Ulmus* species. Johnston and Naiman (1990) documented differential harvesting by beaver among riparian forest species in Minnesota. They found that *Populus tremuloides* decreased in abundance, while *Alnus rugosa* and *Picea glauca* increased, and concluded that the long-term effects on forest succession would be important. In Iowa I have observed extensive cutting by beaver only within a few meters of the river edge. The range of sizes is very variable, but partly gnawed large trees are occasionally left standing, at least temporarily. Also, in areas where large trees are felled, many do not fall completely, but are caught by neighboring trees. I observed that in a stand of *Salix nigra* of some 35 trees, of c. 25 cm DBH and 10 m in height, that had been cut, only about half had reached the ground. *Salix* spp. are often favored by beaver (Aleksiuk 1970; Svendson 1980; Kindschy 1985).

Riparian habitat

While the effects of some animals on the vegetation of riparian areas is important, many reports have documented the importance of riparian habitats for animal species ranging from butterflies (Galiano *et al.* 1985) to grizzly bears (e.g. McClellan 1989). Burdick *et al.* (1989), using the important concept of cumulative impacts, showed that small decreases in the area of bottomland hardwood forest in Louisiana added up to a significant decrease in bird species diversity; Croonquist and Brooks (1991), distinguishing among guilds, found similar results in Pennsylvania. By implication, the losses of other taxa would be parallel. I will not review these reports in detail, nor consider the effect of the riparian area on the aquatic ecosystem of the adjacent stream (e.g. see Hawkins *et al.* 1983; Barton *et al.* 1985; Cummins *et al.* 1989; Gregory *et al.* 1991).

Riparian areas provide a resource for many species that is not available elsewhere in the environment. Brinson *et al.* (1981) summarized some of the key factors making riparian areas valuable for wildlife habitat (Table 4.1). These features include specifically spatial aspects of both the internal and external structures of the riparian landscape. They noted five additional features of importance that would differentiate among wildlife species (Table 4.2). These five factors are interrelated and also tied to the four features listed above. Brinson *et al.* (1981) also provided

Table 4.1 *Features of riparian environments valuable for wildlife habitat*

Woody plant communities
structural variation
woody debris
Surface water and soil moisture
more productivity
Spatial heterogeneity of habitats
edges/ecotones
Corridors
migration and dispersal routes

Table 4.2 *Riparian features which would differentiate among wildlife species*

Vegetation type
mast foods available
nesting and perching sites
Size and shape
interior and edge
Hydrological regime
flooding affects food resources, nesting sites
Adjacent land use
feeding in adjacent areas
Elevation
climate and topography

extensive lists of species diversity and identities of bird species associated with riparian areas, and shorter lists of mammals, reptiles, and amphibians. With these lists they also provided a good bibliography.

In a more geographically and topically focused report, Klimas *et al.* (1981) described the wildlife habitat and species associated with flood regimes in the lower Mississippi River valley. They listed the actual or probable occurrence of species in seven habitat types or landscape elements (marshlands; lakes and ponds; rivers, streams and bayous; swamps and sloughs; wooded bottomland; wooded upland; and grassland, shrubland, and open terrain). For mammals they listed four shrews and voles; 17 bats; two rabbits; 27 rodents (including beaver); 14 carnivores (including wolves, bears, and cougars); two hoofed mammals (deer and boar); the opossum; and the armadillo. For reptiles and amphibians they listed the crocodile; 20 species of turtle; 10 lizards; 36 snakes; 21 salamanders; and 25 frogs and toads. In a list by state they reported some 100 bird species living in seasonal association with riparian habitats. More detailed information on the abundance of some species is given, along with information on the plant communities and a good bibliography.

Work on bird and small mammal populations in riparian areas has found interesting spatial patterns associated with the longitudinal gradients in mountainous areas and patterns of diversity associated with this pattern, the resource, and the relation of the riparian zone to the surrounding uplands (e.g. Knopf 1985; Olson and Knopf 1988). Other studies in this area have focused on the relation between the bird communities and the nature of the vegetation (e.g. Dobrowski 1964; Johnson 1971; Sedgwick and Knopf 1986, 1990). In France, Decamps *et al.* (1987) found that bird species richness was greatest in riparian woodlands, and that the specific mix of patch and matrix was important. These studies indicate that both the spatial relations of the riparian zone and the importance of its hydrological functions are important parameters in the structure and functioning of communities of consumers; specific studies of African animal communities have documented the interaction of consumers with hydrological regime (Sheppe and Osborne 1971; Sheppe 1972). In addition, megaherbivores can have an important impact on riparian vegetation (Dinerstein 1991).

Landscape ecology

While internal structure, or the composition and configuration of tesserae within landscape elements, is not a primary topic of landscape ecology, the importance given at this time to spatial scale in the literature of landscape ecology should alert us to the possibility that spatial configuration and processes of smaller areal units such as tesserae may also exhibit some of the general principles of landscape ecology.

Two major gradients exist in riparian areas. One is a stress gradient related to moisture. The effects of channel dynamics, soil moisture gradients, and the direct and indirect action of flooding combine to create spatial patterns within riparian landscape elements that cause these elements to interact in different ways with the surrounding landscape. In arid areas the riparian area may be the least stressful. In humid regions the riparian area may be the most stressful because of anoxia. In both regions the regime may vary from seasonal drought to temporary anaerobic conditions. The second gradient is that of disturbance resulting from dynamic fluvial geomorphology. It is well established that species can differentiate, or have different niche spaces, along axes related to both these factors. Additional work on productivity and biodiversity is needed along these lines.

The exact formulation of patterns and processes in terms of these gradients are rare in studies of riparian ecology, but examples on which theory is based come from community gradients and differences in distributions which have been described and correlated with environmental variables in many areas in the riparian zone (e.g. Glime and Vitt 1987; Niyama 1987, 1989). Different hydrological regimes create different conditions that are advantageous for establishment and growth of species which differ in their specialization. Thus, on a floodplain, there exist gradients of succession on new substrate, of flood duration and frequency, and of water table depths. The paradigms of individualistic plant distributions, the intermediate disturbance hypothesis, and competition are oriented toward the explanation of these gradients.

Individualistic vs. community explanations

In riparian areas the environmental gradients cannot be duplicated by transects on the ground. Sharp discontinuities and disjunctions in environmental gradients in riparian areas mean that spatially distinct plant communities with sharp boundaries exist in this environment in complex patterns of point bars, cutbanks, abandoned channels, levees, ridges, and swales (cf. Buchholz 1981; Kalliola and Puhakka 1988). These tesserae may develop or may have developed community associations between species, contrary to the individualistic concept, if they have been spatially associated in distinct tesserae through enough time for coevolutionary processes to have operated. Thus, as Westman (1983) demonstrated for some species in Californian coastal sage scrub, strong associations may exist among species, and boundaries along gradients may be strengthened by long-term association in distinct areas. Because the paradigm of species adaptations to an environmental gradient deliberately ignores the actual distribution of species on the landscape, the spatial structure of these distributions and any import it may have has been left for future study. The geomorphological dynamics of the riparian areas, however, may have precluded such long-term spatial associations, as has been demonstrated for the process of edge formation at point bars and cutbanks. Whittaker (1967) considered dynamic processes at single sites as gradients in time, but although he discussed shifting mosaics (Whittaker and Levin 1977), he did not explicitly consider an environmental gradient of which spatial dynamics were a part. I have considered that a general effect of space, i.e. spatial autocorrelation dependent on the relative location of sites and dispersal, could alter species distributions along a strong environmental resource gradient (Malanson 1985*b*). The set of spatial variables such as shape, contagion, and shared boundaries between pairs of elements (Turner 1990) in direct or indirect gradient analysis may provide an informative approach to empirical studies in landscape ecology. While it may seem old-fashioned, empirical data on whether or not these spatial variables are important are certainly needed.

The intermediate disturbance hypothesis

In this hypothesis the frequency and intensity of disturbance at any one place should be distinguished from the landscape-level effects of disturbance. The intermediate disturbance hypothesis was proposed for the processes allowing coexistence at a single site, i.e. processes maintaining alpha diversity. The extension of this hypothesis to propose that disturbances within a landscape in sum are intermediate in frequency and intensity is, however, natural, and the spatially progressive nature of disturbance in riparian areas makes them ideal cases for this extension. What Salo *et al.* (1986) have argued is that an increase in diversity, in the Amazon, due to disturbance is a landscape-level phenomenon, i.e. that fluvial dynamics create different tesserae and increase gamma, or regional, diversity. Studies in riparian areas have not demonstrated the long-term coexistence of species adapted to different degrees of disturbance on a single site which experiences an intermediate level. In contrast, in fire ecology the frequency of fire and the variability of the frequency have been shown to allow the coexistence of species with different niches along the fire frequency gradient (Keeley and Zedler 1978; Malanson 1985*a*). The spatially progressive nature of the disturbance explains the problem for riparian studies. Single locations do not experience a disturbance regime that has a constant mean and variance because of the feedback between disturbance and landform through the geomorphological process. Instead, as on point bars, the site experiences progressively less flooding and sedimentation. Where we see coexistence of species adapted to higher and lower levels of disturbance in this environment is not at a site that is maintained by continuing intermediate disturbance levels, but at a site that is intermediate in its transition from high to low disturbance levels.

Disturbance processes do create different elements and tesserae within riparian landscapes. Where channel dynamics and flooding create disturbances, the species richness may be greater because a greater variety of resource, disturbance, and regenerating patches can coexist within the landscape. Distinct plant communities exist on point bars, in abandoned channels, at cutbank edges, and in extensive floodplain interiors. The diversity of these several communities is maintained by active fluvial processes. Differences in both life history and physiological traits among species separates them among tesserae. The disturbance expands the range of environmental gradients available. In contrast to some other environments, where the disturbance primarily allows ruderal or fugitive species to occupy transient sites, in riparian areas the expanded gradients include very long-term changes in site conditions, such as in abandoned channels, as well as transient sites such as point bars and channel bars. At the landscape level, therefore, the range of disturbances produces a range of spatial and temporal dynamics, i.e. the shifting mosaic of complex environmental gradients with specific spatial characteristics.

Competition

The distribution of species on environmental gradients and the pattern of species diversity on disturbance gradients must be seen in relation to the development of a theory of competition. Competition structures the arrangement of species along environmental gradients. It affects the overlap of species along multiple gradients and thus alpha diversity, and consequently it determines the number of species that can be packed along the total lengths of all such gradients or the total volume of the niche hyperspace, and thus gamma diversity for the landscape.

At present it appears that the concept of centrifugal organization (cf. Rosenzweig and Abramsky 1986; Keddy and MacLellan 1990) offers the best avenue for elucidation of competition, disturbance, and environmental gradients, if spatial and temporal variation can be included in the model. At present, the model, although it includes disturbance, is rather static. Long-term changes that may underlie recurrent disturbances are not included in the model, and it is not clear how a community might change through time toward any other combination of species except by migrating toward the center of the diagram in the absence of competition. The importance of relative position around the circle is not clear. Also, effects of area, isolation, shape, and related features of landscapes are not included. It may be necessary to expand the visual, and perhaps in some sense the fundamental, presentation of the model in order to accommodate time and space *per se*. The centrifugal model, as diagrammed by Keddy and MacLellan (1990) is an extension of Grime's (1979) (also aspatial) triangular model, which embodies a relation between competition and disturbance not specified by Grime (1979). A way to add time and space to a centrifugal model might be to expand it into a spherical or at least hemispherical model. This expansion adds a vertical dimension which may be thought of as having the least spatial restrictions at the center, where competitive species dominate in the region of greatest productivity. On vectors away from the center and with increasing height above the plane, spatial isolation increases and size decreases, and the importance of disturbances and founder effects increases and fugitive species can persist.

In most riparian areas, removal experiments would be difficult where the dominant species are large trees. It may be possible, however, to find situations analogous to removal experiments in natural and human-influenced processes. Even without the actual removal of individuals or species, when the environment changes drastically, leaving species outside their normal range, it may be possible to determine the effects of competition through analysis of disequilibrium. If species can continue to live and reproduce in an area after the environment has changed (e.g. change in flood frequency caused by an upstream dam), even if temporarily until their competitors arrive, then a competitive hierarchy will be demonstrated. If, however, species die out following such a change before the arrival of their competitors, then distinct niche differentiation will be demonstrated. Areas where the vegetation is fragmented or naturally discontinuous, and thus where isolation can control the rate of immigration of competitors, may prove useful in such studies. Bravard *et al.* (1986) have provided what may be the best case in riparian landscapes for constructing such a model. Their conceptualization includes a major temporal component; what they refer to as their spatial component is differences in the physical environment and impacts that characterize individual sites. Their model does not include quantifiable spatial landscape features, but this research team has excellent maps at its disposal and could easily derive such variables in a future model development.

Conclusions

In riparian landscapes the internal structure provides many opportunities for investigation of the relations of species distributions on environmental gradients, disturbance, and competition. Studies of the spatial characteristics of individual tesserae, such as abandoned channels, cutbanks, or point bars, will serve as a basis for an integration of such sites along environmental gradients in a way that gradient studies in the past, which did not examine spatial dimensions of the tesserae they crossed, could not. In some riparian areas the role of *Salix* species as fugitives primarily inhabiting point bars may indicate the way in which a dynamic mosaic might play a role in alpha and gamma diversity. On larger, more accessible point bars the replacement of *Salix* by *Populus* and other species may be faster, and the ultimate maintenance of several *Salix* species in a riparian landscape may depend on the spatial dimension.

The internal structure of the riparian zone provides a setting for the examination of the principles of landscape ecology at a scale usually considered to be within a single landscape element. By stepping to a finer scale some of the logistic problems and problems of study design (i.e. what is replicative and independent while maintaining control of historical variation) can be reduced. The study of spatial pattern and processes of tesserae can perhaps serve as a basis for studies at the coarser scale, and provide insights into the effects of scale *per se*.

5 · *Cascades of material and energy*

Along the river and on the islands, on the edge of the dyke and far away in the distance, one sees only poplars. In my mind there is a strangely intimate relationship, a strangely indefinable resemblance, between a landscape made up of poplars and a tragedy written in Alexandrines. The poplar, like the Alexandrine, is one of the classic forms of boredom. *(Victor Hugo)*

Functions in riparian forests operate through cascades of water, sediment, nutrients, contaminants, and carbon. It is in these cascades that the lateral flow of water dominates many of the processes in riparian landscape elements, and these are the cascades that primarily link the riparian elements to the other elements in the landscape. The primary cascade of water in the riparian zone is over-bank flooding, but precipitation, groundwater flow and evapotranspiration are also important. The role of the riparian zone in the storage of water is of critical importance. Sediment and nutrient cascades are directly linked to the cascade of water. The riparian zone may be an area of net erosion or deposition, or, for some time, an area of transport where sediment inputs are balanced by outputs. The sediments moving into or out of riparian areas are often sources of nutrients for the biota and may also be sources of toxic materials. Cascades of organic energy, beginning with photosynthesis, are also important in riparian ecosystems. In forested areas large woody debris from fallen trees interacts with water and sediment cascades, and the processing and movement of leaf litter is also affected by flooding. In examining the interaction of these cascades in terms of landscape ecology, areas of intense study contrast with areas that have received scant attention.

Chauvet and Decamps (1989) provided a brief outline of the factors to consider: the interaction of hydrology, nutrient fluxes, carbon fluxes, and human activity. These flows will be examined in relation to the roles of the longitudinal and transverse aspects of the landscape structure and of the internal structure of the riparian element. Pinay *et al.* (1990) identified many of the areas of riparian function, e.g. water and nutrient fluxes, and did so having identified longitudinal and transverse structure as important factors. They considered, however, that the riparian zone is primarily a transverse ecotone between the terrestrial and aquatic ecosystems, while longitudinal ecotones occur in-stream. Chauvet and Decamps (1989) and Pinay *et al.* (1990) alluded to the resource spiraling concept. This concept was developed in lotic ecology as a representation of the in-stream processing of nutrients, with tighter spirals signifying longer residence times in a reach for a given element or molecule and wider spirals signifying a more open system with higher rates of flux and less storage (Newbold *et al.* 1981; Elwood *et al.* 1983). The French group expanded this concept to the riparian corridor as a whole. This framework provides some common ground for consideration of cascades of material and energy. The flood pulse concept goes beyond this framework by recognizing the nonequilibrium interactions between hydrology, geomorphology, and ecology (Junk *et al.* 1989).

Cascades of water

Cascades of water in the riparian zone can be examined in three related areas: over-bank flooding from the river, storages on the floodplain and subsurface flow, and evapotranspiration. The three are connected as factors in the local hydrological cycle. Over-bank flooding may be a primary source of water for soil moisture and groundwater in the riparian zone, and thus also for evapotranspiration. The short- and long-term storage of water in the riparian zone, and its use by plants, can also affect the hydrological regime of the river and its effects on the downstream riparian area.

Over-bank flooding

Flood events are important cascades that affect the structure of riparian forests. The specific responses of plant species to the frequency and duration of flooding have been reviewed above. The riparian forests also influence the flood regime, both in the adjacent stream and especially downstream. It has been well demonstrated that wetlands, including riparian forests, by their topographic form, are storages for flood waters. Novitzki (1978) documented the difference in spring peak runoffs in Wisconsin between basins with and without lakes and wetlands. This function of flat areas is well understood, but the details needed to precisely calculate the movement of water through a floodplain forest are lacking. The over-bank flow which enters storage on the floodplain

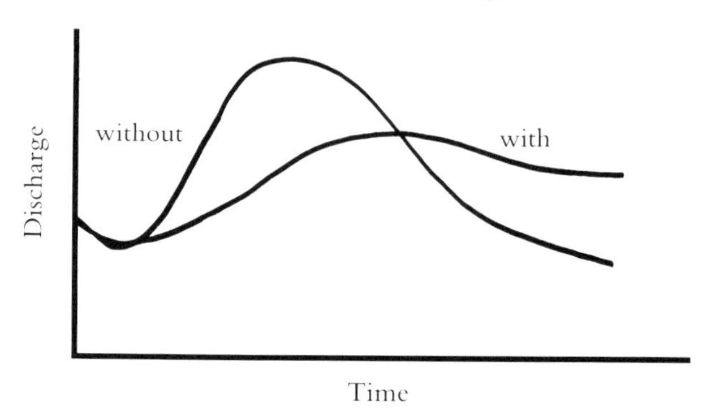

Figure 5.1 Hydrographs downstream of an area of over-bank flooding have lower peaks but longer periods of flood than they would without this process.

reduces the channel flow and the height and energy of the flood wave. At the upstream site the energy of the moving water is absorbed by impact on the trees as the forest is flooded. Thus downstream flooding does not reach as high nor is it as intense, but it may, however, be of longer duration (Figure 5.1). Therefore, the absorption of water at one place upstream will affect the process and structure downstream. The floodplain creates tighter spirals locally and effectively tightens downstream spirals by dissipating flood energy.

Two approaches which potentially include interaction with riparian vegetation have been used to assess flood flows. Hydrological simulations have been used to address the routing of a flood wave down a river channel, and botanical evidence has been used to assess the frequency and magnitude of past flood events.

Simulation models

Simulation models of hydrological flows cover a wide range of techniques. At the simple end of the gradient, models are statistical, lumped parameter models for a spatially averaged basin. At the more complex end, models are process-based and spatially explicit. It is these latter models which are of interest from the perspective of landscape ecology. One particular model represents the developments of process-based, spatially explicit hydrological models. The model SHE (Systeme Hydrologique Européen) was developed by a team of European hydrologists (Abbott *et al.* 1986*a*,*b*). This model divides a basin into rectangular cells, the dimensions of which can vary. The dimensions must be changed for an entire row or column of cells, however, and fine detail in the center of a basin must be represented by attendant detail at the edges of the same coordinates. The cells represent land areas and so the river channel is assumed to follow cell boundaries, thus never being truly curvilinear. WATBAL uses an interesting spatial technique to include different levels of resolution in different areas of the basin and for different vertical hydrological units (Knudsen *et al.* 1986). It specifically includes a riparian zone. None of these basin models includes over-bank flow, however, and Anderson and Rogers (1987), in their review of basin hydrological models, and Band and Wood (1988) in their emphasis on including terrain information in models, did not address this question as a potential research direction. Once water from surface or subsurface flow enters the channel, it remains there. The projection of over-bank flooding requires specific flood routing equations.

Flood routing equations have been developed for engineering purposes. These equations show either velocity or discharge as a function of slope, flow depth, and friction. Ponce (1987) presented an advanced formulation to include whole catchment dynamics, but when such flows move over a floodplain their calculation becomes even more complex (Rajaratnam and Ahmadi 1981; Baird and Anderson 1992). Pasche and Rouve (1985) developed a model that incorporated the effects of a riverbank populated by trees. Using extensive theoretical calculations and experiments with a flume, they found that field measures could be reasonably approximated by an extensive set of equations for which a simple algorithm exists. They divided the transverse dimension into four parts, two areas, of floodplain flow and main channel flow, each subdivided by whether or not they were influenced by the other: in mid-channel is main channel flow uninfluenced by floodplain flow (but by bed roughness); in channel but closer to the bank is main channel flow influenced by floodplain flow (friction of vegetation on the floodplain is translated out into the channel); floodplain flow influenced by main channel flow is in the vegetated areas adjacent to the river where both effects are strong and the energy of main channel flow is dissipated; floodplain flow not influenced by main channel flow is in the vegetated floodplain at some distance from the river, before reaching which the energy from the main channel flow has already been dissipated. To this transverse structure they applied the Darcy–Weisbach law as the equation governing momentum. Krishnappan and Lau (1986) modeled floodplain flow in two dimensions, and Sengupta *et al.* (1986) modeled the effects of vegetation on water flow across the Everglades; they included effects of evapotranspiration as well as eddy diffusion in three dimensions. Li and Shen (1973) modeled the effect of tall vegetation on floodplain flows, and Smith *et al.* (1990) modeled flows and did experimental work on the topic. These types of models will need to be generalized and made part of basin hydrological models. But they too have notable simplifications, specifically that the topography and vegetation of the floodplain are homogeneous. Hammer and Kadlec (1986) included vegetation as a factor affecting their simulation, but their topographic field was very simple compared with riparian areas with multiple channels. They were able to account for multiple flow directions, however, and this model may be a basis for further development.

At a landscape scale Ogawa and Male (1986) simulated the role that wetlands could play in minimizing downstream flooding and associated damages. Their simulation accounted for partial filling of wetlands and for their position within a drainage basin. They found that small changes in wetland area had minimal effects on flood peaks, and that wetlands on the larger streams were more effective than those on lower-order streams. This type of modeling will be increasingly important as a guide to managers as well as for the study of the interactions of geomorphological and ecological processes. Models focused on sediment transport will be discussed below.

For applications in other situations a number of simplifications have been used. To calculate flood flows in a narrow canyon from flows at a downstream gauge, Malanson and Kay (1980) used the Manning equation:

$$V = S_f^{1/2} R^{2/3} / n,$$

where V is the velocity, S_f is the energy slope, R is the hydraulic radius, and n is a roughness factor. In this case simplifying assumptions about channel geometry were made possible because the river occupied the entire bottom of a canyon with vertical walls. We calculated flash flood frequency as an indicator of damage done to plant communities (Figure 5.2); this calculation indirectly incorporated the effects of the kinetic energy of the flow on the vegetation. Smith *et al.* (1990) studied the effects of vegetation on overland flows, reviewing earlier work and conducting new experiments in sorghum, maize, barley, and wheat, and concluded that the Manning equation was inappropriate for such flows. Arcement and Schneider (1989), however, specifically calculate Manning's roughness coefficient for vegetated floodplains, including those with dense vegetation of trees, shrubs and vines. In their simulation of flood effects on internal structure, Pearlstine *et al.* (1985), discussed

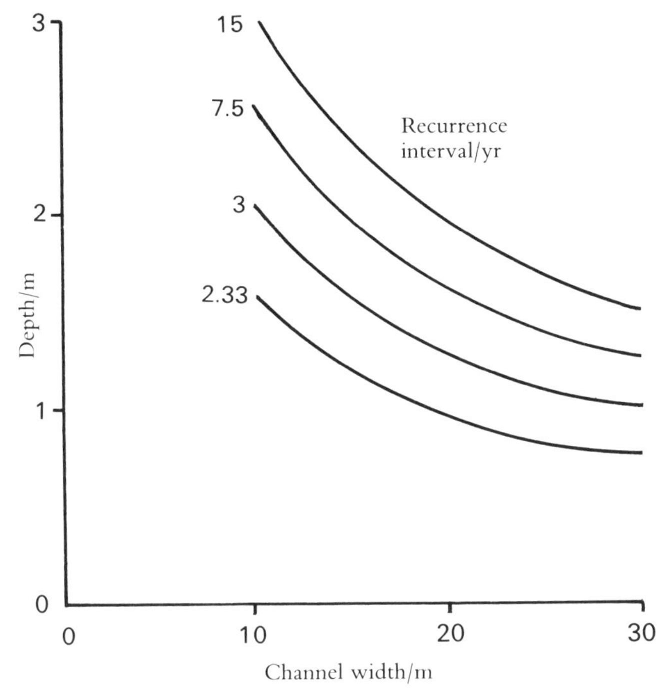

Figure 5.2 Family of flood curves calculated for the Virgin River in the Narrows, Zion National Park, Utah, in order to estimate the return interval for floods of hanging gardens.

above, used a linear regression to interpolate the height of flooding in a riparian forest between two stage gauges. In a river with a low gradient and without major obstructions this technique is satisfactory. A major problem will arise, however, in sloughs or other depressions where standing water may remain, depending on soil moisture and infiltration rate, long after the river has receded. Models of water flow through wetlands or other generally flat areas may be applicable to such problems (cf. Young *et al.* (1972) for a simple approach to this problem using a physical simulation built from five watertight plastic boxes).

Dendrohydrology

The most common technique using botanical evidence to assess the frequency and magnitude of floods is dendrochronology, and the term dendrohydrology has been generated in this context. Tree-rings can be used to study flood events in two ways: by examining ring widths, where growth is expected to be related to either drought or precipitation which would affect river flows or where growth is affected by the conditions of flooding itself, i.e. anoxia; or secondly through the examination of specific features such as reaction wood, where the force of the flood tilts or topples a tree and it responds with uneven productivity, or response to toppling and burial (Shroder and Butler 1987).

At one level tree-rings have been used simply to assess annual runoff. Trees from any site within the basin might suffice (Phipps 1972, 1983) where growth is simply correlated with basin precipitation. Stockton (1975) used tree-rings to reconstruct river flow. Most notably Stockton and Fritts (1971) documented that the years of record on which the discharge of the Colorado was legally divided among the riparian states was an unusually high flow period. Landwehr and Matalas (1986) reported that statistical considerations in this methodology may need to be more sophisticated than they had been in earlier work.

Other studies have focused on the impact of high water on tree growth and have used this information to reconstruct river discharge records. Particularly narrow rings in riparian forest indicate a reduction in growth due to flooding (cf. Wendland and Watson-Stegner 1983; Bowers *et al.* 1985). Stockton and Fritts (1973) calculated the long-term flood regime for the Peace River, Canada, using this technique. More detailed analyses of the annual growth of each ring show that in some cases a flood can disrupt the pattern of growth by causing the onset of new vessels, typical of wood produced early in the growing season, which may then appear as a less distinct ring (Yanosky 1982*a*,*b*, 1983). This phenomenon probably happens when the growth is radically interrupted, such as when a flood might defoliate a sapling.

Sigafoos (1964; see also Sigafoos and Sigafoos 1966) developed the methodology of using reaction wood as an indicator of flooding and pioneered the use of tree-rings to document the growth of sprouts or vertical branches from downed trees and tree growth when parts have been buried by alluvium (my graduate students refer to such trees as 'Sigafoos trees'). LaMarche (1966) and Laing and Stockton (1976) used tree-ring evidence related to the exposure of roots by floods in the Colorado Plateau region of the southwestern USA. Tree-rings also reveal more direct evidence of damage caused by flooding. Butler (1979) used the onset of reaction wood as an indicator of major flood years on the Mackenzie River, Canada. Begin and Lavoie (1988), Gottesfeld and Johnson Gottesfeld (1990), and Payette and Delwaide (1991) used dendrochronological methods to document the floodplain hydrology of rivers in Canada, and Fagot *et al.* (1989) used tree rings to reconstruct the population changes in *Populus nigra* and so infer the fluvial metamorphosis of the Ain in France. All of these efforts are generally reliable indicators of flood regime although an adequate sample size is needed to avoid documenting other events affecting one or a few trees (Butler *et al.* 1987). Malanson and Butler (1992) failed, however, to find rings from trees at the cutbank edge of a terrace to be useable indicators of flood events because of individualistic responses and the contradictory effects of release from competition as other trees toppled during erosion events.

The annual growth of tree-rings is, however, sensitive to a variety of factors. In riparian areas, the effort to assess hydrological effects may be confounded by trends in climatological parameters (Bowers *et al.* 1985). In Iowa we observed that temperature and precipitation also affected ring widths in *Salix nigra*. An important factor that is difficult to resolve is the effect of competition by neighboring trees; dendroclimatological studies have often used trees in isolated sites which would also produce a sensitive response in ring width to variations in climate, but this approach may not be possible in many riparian areas. Tree-ring studies may prove valuable in a landscape ecology approach to the cascade of water in riparian areas. Dendrochronology can help reconstruct the spatial distribution of past responses to environment and may be able to be used in the generation of a model of landscape reproduction, or tree-rings may be used to test hypotheses if the relations which they represent are clear.

Storage on the floodplain

Ponding

During and after an over-bank flooding event much of the water will move downstream, but a portion will be used to replenish soil moisture up to the point of saturation, and some will remain in sloughs, abandoned channels, and other depressions from which it will subsequently evaporate or infiltrate. In some areas this water may be the primary way in which permanent ponds are maintained. These storages are the most significant ways in which a spiral of the water cascade would be greatly tightened over a long period, and the ecological effects are also important. Mitsch *et al.* (1979*a*) observed the short-term effects of floodplain storage on the flood hydrograph of a river in Illinois. It is difficult to project the long-term effects, however, because of the complexity of the microtopography and soil hydraulic properties (cf.

Shirohammadi *et al.* 1986). It has been estimated that a large proportion of the runoff in some areas with gravel bed rivers is subsurface, beneath both the riparian area and the river channel. In other areas permanent floodplain ponds may be maintained by over-bank flooding. I have not found a hydrologic budget for such a pond that would test this idea. Williams (1987), in an analysis of temporary ponds including floodplain ponds, cites no such studies; although he reported on the process in general and presented data on pond level for a small pond, he did not actually quantify rates of evaporation and infiltration. Also, Rostan *et al.* (1987) have found that in abandoned channels in France the sediment characteristics indicate differences between ponds that receive regular inputs of river water and those maintained by groundwater flow.

Groundwater monitoring

The flux of groundwater has been modeled for many conditions. It seems rather simple to apply Darcy's law with a known hydraulic gradient and conductivity and thus calculate flow. The problem is much more complex, however, in reality. General conceptual difficulties arise when non-saturated states may exist and interact with saturated flow, and, significantly, complex spatial patterns of sediments in riparian floodplains lead to complex mixtures of hydraulic gradients and conductivities. Empirical studies have attempted to actually measure flows, however. Ruddy (1989) used piezometers (i.e. wells) (Figure 5.3) to try to determine the direction of flow, and followed up that step with a detailed study of one area using a hydraulic potentiomanometer. She found that at the upstream end of the riparian wetland the gradient was away from the stream, while at the lower end it was the reverse. Also using piezometers, Riegel *et al.* (1991) found a consistent gradient toward the stream in a riparian montane meadow, while Sundeen *et al.* (1989) found little connection between valley bottom wetlands and the main stream channels in a subalpine area. Ulmer (1990) found that soil type greatly affected the degree of influence of a pooled section of the Tennessee–Tombigbee canal on the neighboring floodplain; the effect was noticed from 300 to 500 m, depending on season.

Other studies in hydrology have attempted to relate groundwater flux to precipitation and/or river levels. Bell and Johnson (1974) found a strong association between river stage and level of the water table near the river (25 m) which decreased markedly over a gradient of 200 m and an elevation change of c. 1 m. Grannemann and Sharp (1979) documented four types of response in groundwater responses to changes in

Figure 5.3 Shallow wells with openings to the soil can be used to measure the height of the water table. In this case standing water indicates that the water table is effectively above the surface.

the stage of the Missouri River. Doty *et al.* (1984) showed a relation between groundwater levels and river stage, which decreased with distance to the river, that they were able to modify experimentally. Sophocleous *et al.* (1988) also reported complex responses in groundwater to streamflow. Studies of the relation of riparian vegetation to groundwater, discussed in Chapter 4, which monitor groundwater include those of Kondolf *et al.* (1987), Space (1989), and Girault (1990). These studies indicate that the particular characteristics of sites make quantitative generalizations impracticable.

The subsurface flow of water in a drainage basin has much tighter spiraling than the surface flows, but the two are connected, especially in the riparian area. The nature of these spirals and their ecological consequences need more study. It is clear from work reported in Chapter 4 that groundwater levels are important, and groundwater fluxes are the basis for this system. While groundwater flow may be a minor part of the total cascade of water in a riparian area, it is notable that it is likely to differ greatly among regions depending on substrate.

Evapotranspiration

Riparian environments also affect the water balance. In absorbing energy by evaporation they add moisture to the air (cf. Munro 1979; Zidek 1988). The effect of the stream itself on the local humidity may be small (Ellenberg 1987), but this may depend on its size and velocity. Thus it is likely that humidity is high within (this effect is experienced directly by researchers trying to work in this environment!) and downwind of riparian forests (Vasicek and Pivec 1991*a*,*b*). In climatology this phenomenon is referred to as the 'oasis effect', and has been recorded primarily for bodies of open water (Oke 1987). If the microclimate around a riparian area is examined, certain relations may be hypothesized. First, if one assumes wind from one direction and a particular topographic form and surrounding land use, then general principles of environmental physics and microclimatology can be applied. For the case shown in Figure 5.4, one could postulate first the oasis effect, shown as the change in temperature and humidity relative to the riparian area, and also a 'clothesline effect' where the windward edge of the riparian area itself is modified by the heat and humidity of the upwind environment. Because of the heat capacity of the greater amount of water stored in the riparian area, in open water, soil, and vegetation, it will also respond more slowly to changes in the regional temperature following frontal passage. This inertia may also be found during periods of long-term climatic change where a riparian area might alter its water budget more and its temperature less as the regional mean temperatures change, but this effect has not been investigated even in models.

Through the process of evapotranspiration, vegetation removes moisture from the local environment, and riparian areas may be exemplary cases (Hicks *et al.* 1991). This process is most noticeable in the southwestern USA where phreatophytes can significantly alter the flow of streams. Robinson (1958) presented basic information on phreato-

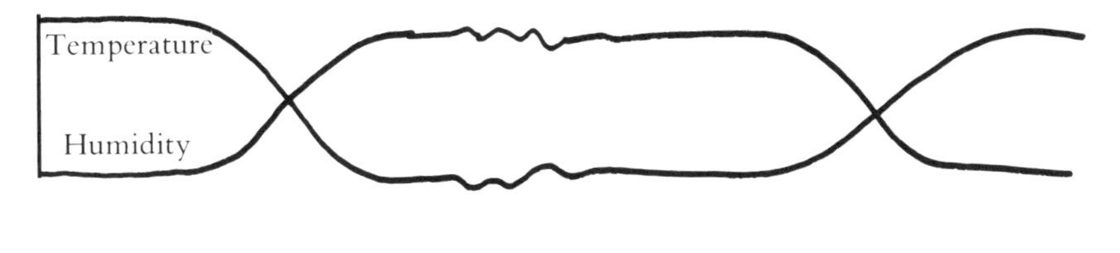

Figure 5.4 While any change in vegetation can modify the microclimate, riparian landscapes may be more powerful in altering humidity and temperature levels than areas with less accessible water.

phytes. He provided definitions and outlined the related problems of groundwater depletion and the influence of phreatophytes on siltation and flooding. This latter topic has recently been reemphasized by Graf (1985). Robinson (1965) concentrated on documenting the problems associated with the Eurasian invading phreatophyte, *Tamarix chinensis* (formerly *pentandra*; salt cedar). Horton (1972, 1973, 1977) reviewed the literature related to the process. A number of studies have attempted to measure the actual amount of water transpired by riparian vegetation on sites in the the southwestern USA (e.g. Sebenik and Thames 1967; van Hylckama 1974; Gay and Hartman 1982; Patten 1984), but Graf (1985) noted that for a variety of reasons, including wide variations in the estimations of phreatophyte evapotranspiration (e.g. Davenport *et al.* 1982), ranging from 34 to 280 cm of water per year for *Tamarix* sp., the most important phreatophyte (but Gay (1985) reported an estimate of 12 mm day^{-1} in July and 1727 mm yr^{-1}), the actual importance over an entire basin could not be known. Dawdy (1988), however, reported basin-wide effects of consumptive water use by natural vegetation in the Central Valley of California on the order of 4.5×10^{10} to 6.0×10^{10} m^3 yr^{-1}.

Other studies on the effects of riparian evapotranspiration on water balance have been undertaken in the Soviet Union. Popov (1985; Popov and Cherenkov 1985) linked together studies of soil moisture and groundwater with studies of microclimate in mixed deciduous–coniferous forests. This more comprehensive type of study is needed to link together the effects of floods and the level of the water table. Eckersten

(1986) simulated transpiration by *Salix*, and found that it could be predicted by relatively simple measures of air temperature and humidity.

Evapotranspiration creates an alternative pathway for part of the cascade of water and also creates moisture gradients in the soil that affect subsurface flows and spiraling. Overall, riparian hydrology is remarkably poorly understood. Given the intensive work on wetland ecosystem processes, including forested floodplains, one would expect that models would be more advanced, but none combine precipitation, over-bank flow, groundwater flow, standing water, and evapotranspiration. It may be a general feature of ecosystem studies that they are difficult to apply at a landscape level.

Sediment and nutrients

Two aspects of the role of riparian vegetation in the cascade of suspended and dissolved material have been considered in terms of sources vs. sinks: water quality and geomorphology. The aspect of water quality has been considered for the hypotheses that riparian zones intercept material from uplands before it reaches the stream, and that they serve as a sink for the deposition of materials from the stream, or that they may at other times be a source. The view of fluvial geomorphology is focused on the role of riparian environments in stabilizing the banks of eroding channels and in functioning in the alluvial process; their role at any one point varies in time. These cascades are also directly related to the ecosystems along and in streams. It is in the area related to water quality, specifically regarding nutrient inputs to aquatic ecosystems, that the spiraling concept was developed and to which the landscape-level expansion of Chauvet and Decamps (1989) and Pinay *et al.* (1990) is most germane.

Water quality

Riparian areas may act to absorb nutrients, and thus act as filters and improve water quality both as a linear barrier between upland surfaces and the stream and during over-bank flooding when nutrients already in the stream may be filtered (they may also contribute nutrients) (Finlayson 1991). This filter effect is the major tightening of the nutrient or sediment spiral in the downstream direction. Again, surface and subsurface flows and effects need to be distinguished, and in this area the distinction has been made and its importance is clear.

Studies of the role of wetland vegetation in absorbing nutrients have been reviewed. Kadlec and Kadlec (1978) summarized much of the pertinent information for wetlands in general. They noted that the key factors to understand were the inputs, outputs and storages of both water and the elements of interest. Klopatek (1978) examined a riparian system along these lines, but like most of this work he concentrated on herbaceous vegetation. A considerable amount of information on wetlands themselves is still needed to evaluate them (Stuber and Sather 1986), but the problem of intercepting and absorbing major inputs of agricultural runoff and subsurface flow must take into account the borders of every stream, not only those areas that are flooded most of the time. Although almost all studies are concerned with agriculture, Brown (1988) and McCullough (1988) found that in general wider riparian zones improved water quality in areas of forest harvest, and Osborne and Wiley (1988), using a geographic information system to define 100 foot (33 m) buffer zones around a stream network, found that urban land uses also played an important role. Although this quantification of landscape structure for the analysis is excellent, they did not explain their regression methodology in enough detail to determine if it was appropriate for their data set, which had a notable lack of independence among cases and almost certain high multicollinearity among variables. Although the interaction of river flood waters and groundwater has been demonstrated in the context of the flux of pollutants (Carbiener and Tremolieres 1990), most studies of the filtering effects of riparian vegetation concentrate on either surface or subsurface effects, not both. In detailed studies of nutrients, however, Lowrance *et al.* (1984*a*,*b*,*c*) quantified the role of riparian areas as sinks for nutrients in both phreatic and surface flow.

Surface filter effects

Karr and Schlosser (1978) presented an early elucidation of the role that riparian vegetation can play at the interface between agricultural ecosystems and aquatic resources. They argued that forested channel banks and natural, as opposed to straightened, channels would reduce the flux of sediment and nutrients to a stream as well as reduce the incidence of higher water temperatures detrimental to some stream fauna. This report has been followed by a wide variety of investigations which seek to understand the processes by which riparian vegetation has an effect on water quality.

Riparian areas may absorb sediment and nutrients from the stream during over-bank flows, while at other times riparian forests may act as sources of material for streams. Mitsch *et al.* (1979*a*) reported that an alluvial swamp forest absorbed considerable levels of sediment during over-bank flows of an Illinois river. Flood events may in some cases erode material, but deposition seems to outweigh erosion in this case. Schlosser and Karr (1981*a*) reported changes in water quality during base flow statistically associated with forested riparian areas. While variations in flow and in-stream organic production were related, riparian vegetation can apparently control inputs of sediment and adsorbed phosphorus, but contribute phosphorus through leaf litter. Schlosser and Karr (1981*b*) also reported that use of the Universal Soil Loss Equation (Wischmeier and Smith 1965) as an independent variable to statistically predict suspended solids, turbidity and phosphorus was affected by the presence or absence of riparian vegetation and channel substrate stability. Omernik *et al.* (1981) examined the effects of the relative location of farmland and forest in a basin as opposed to the simple proportion of basin in each land use and found no effect on water quality. It appeared that an area of upland forest reduced sediment output as much as an equal area of riparian buffer strip. The authors suggested that over the long term riparian vegetation may not act as a sink for nutrients, although they interjected several caveats concerning their study, most notably that they did not differentiate effects of grazing.

Lowrance *et al.* (1983, 1984*c*, 1985) also described the potential importance of riparian forest areas for improving water quality. Lowrance *et al.* (1986) measured long-term storage of sediment, eroded from uplands, in riparian areas. They found that basin sediment yield was less than expected from upland erosion. The riparian zone was the obvious store. They found spatial variation in deposition associated with the edges of agricultural fields, where small berms developed, and with stream order. While they had hypothesized that deposition would increase with stream order, they found that deposition was low on higher-order streams, perhaps owing to the proximity of agricultural fields to the lowest-order streams and to greater sediment loads in higher-order streams during storm events. Lowrance *et al.* (1988) used cesium-137 to compare sediment deposition in a riparian forest with erosion in an adjacent upland field. They found that deposition had exceeded erosion by a factor of 2, and concluded that the riparian forest served as a sink for sediment delivered in over-bank floods from upstream locations as well as from the adjacent field.

Dillaha *et al.* (1989) and Magette *et al.* (1989) used rainfall simulators on small experimental plots to measure the effect of small grass filters on sediment and nutrients in surface runoff. Dillaha *et al.* (1989) found good retention of suspended material but in their case the filter served as an additional source for soluble nutrients. At higher flow rates the filters were less effective. Magette *et al.* (1989) found variable effectiveness in reducing nitrogen and phosphorus in filters of 4.6 and 9.2 m, but that effectiveness decreased with successive runoff events. Both of these studies concluded that such grass filter strips could not be the primary on-farm strategy for reducing sediment and nutrient losses to streams.

While most studies have emphasized the role of the riparian filter in terms of its location between upland sources and the river, retention of overbank flood waters also leads to filtering (Phillips 1989*a,b,c*). Kleiss *et al.* (1989) examined a longitudinal filter of the Cache River, Arkansas. Water was sampled from the main stem above and below an area, c. 25 km in length, with extensive riparian floodplain forests. When the river was high enough to flood the forests, nutrients were retained. From this longitudinal perspective, the hydrological regime is most important, as opposed to the ecological dynamics found for lateral filters. Sanchez-Perez *et al.* (1991) traced nitrogen and phosphorus fates after flooding along the Rhine; both elements were retained in the plant–soil system and differences noted among plant communities because of the differences in the nitrification and denitrification and phosphorus transformation processes in the associated soils. Beaver activity also contributes to the landscape ecology of a basin and to its longitudinal processing of water and nutrients. Naiman and Melillo (1984), Francis *et al.* (1985), and Naiman *et al.* (1986) documented the fixation of nitrogen in beaver-controlled reaches of streams. They concluded that the biogeochemical processes in natural streams with beaver must be conceptualized differently than for those without.

While the spiraling concept in lotic ecosystems was proposed for the longitudinal gradient, it is clear that if the riparian landscape is to be included in a spiraling model, then the transverse structure of the landscape and transverse spirals will need to be considered. A new formulation would divide the riparian element into connected two-dimensional tesserae, each of which could spiral material in multiple directions. The time steps of the model will also need to be considered carefully: as discussed below for sediment, intermediate storage in the riparian zone does not necessarily mean permanent storage. Vanek (1991) found that the 'riparian' zone of a lake was contributing phosphorus, which had been stored there after earlier upland erosion and runoff, to the waters, and Peterson and Rolfe (1982*c*) found significant rates of uptake of nitrogen, phosphorus, and potassium by herbaceous floodplain species which hold nutrients during times when they might otherwise be lost, and then, because they also decompose quickly, also act as a source.

Subsurface filter effects

Schlosser and Karr (1981*a*) examined the function of riparian vegetation in absorbing nutrients during baseflow, and found that the stability of flow improved the ability of the vegetation to act as a sink. Lowrance *et al.* (1984*a*,*b*,*c*) noted that in addition to trapping sediment, riparian zones could filter nitrogen through uptake by the vegetation and through denitrification. Using wells on transects from uplands through riparian forest, to the stream edge, they computed mass balances and computed storages and outputs for the riparian ecosystem. In addition to effects of nutrient inputs at different times in the agricultural cycle and different rates of processing them in different seasons, they found that riparian forests were a sink for phreatic nutrients and that nutrients were higher in artificial drainage waters which bypassed the natural riparian ecosystem.

In North Carolina Jacobs and Gilliam (1985*a*,*b*), Gilliam *et al.* (1986) and Cooper *et al.* (1986, 1987) have documented closely the processes by which nitrogen and phosphorus may be trapped in riparian forests. Cooper and Gilliam (1987) and Cooper *et al.* (1987) documented the filtering of phosphorus and sediment by riparian ecosystems. They found that the riparian zone was a sink for both, and that differences in the spatial distribution of both occurred within the riparian zones.

Peterjohn and Correll (1984, 1986) measured nutrient fluxes from both an agricultural field and a buffer strip of riparian vegetation between the field and a stream. They studied the absorption of nitrogen and phosphorus in a Maryland riparian forest which occupied about 30% of the small basin, or about 50 m of width perpendicular to the stream. They found that the riparian buffer zone was able to store nitrogen and phosphorus but that outputs from the riparian zone to the stream in groundwater and leaf litter could be important. In addition to sequestering nutrients, the riparian zone also had a positive effect on a steady stream flow and reduced the acidity of water moving through it. They did not measure denitrification.

Fail *et al.* (1987) measured differences in the growth and nutrient concentrations of riparian forest trees in different spatial relations to agricultural areas and concluded that they were capable of uptake of considerable masses of nitrogen and phosphorous. They appear, however, to have misinterpreted their results. Except for one riparian forest site affected by nutrient runoff from a pigpen, the nutrient concentrations of the two riparian test sites and the two control sites do not appear to be significantly different. The problem lies in the study design: the difference between the control and test sites is not whether they are riparian or not, but whether they are immediately adjacent to an agricultural field or separated from it by a grass field. This study does not test the effectiveness of riparian sites in absorbing nutrients; it tests the effectiveness of the grass fields, which appears to be minimal (see Dillaha *et al.* 1989; Magette *et al.* 1989, cited above).

Licht and Schnoor (1989) conducted an interesting experiment in the effectiveness of a riparian filter strip of trees. Beginning with unvegetated straightened first-order channel, they planted strips of *Populus* along the stream using cuttings of 30 and 150 cm. The trees grew rapidly and produced extensive root systems. Samples of nitrate and total organic carbon from wells in the study area and of nitrogen concentrations in the trees indicated that the trees sequestered nitrogen and that the additional carbon provided at depth in the soil increased the rate of denitrification. Licht (1990) reported more extensive results. The deeper planted cuttings performed better than the shallow ones in some respects. Both lowered nitrate and total nitrogen in the soils and groundwater in the plots, but the effect was most notable for the shallow soils of the deep planted cuttings. Organic carbon was increased in the middle-level soil (1 m) on these same plots. In terms of groundwater, the treatment had a notable effect within the plot, but not downgrade from it. This lack of effect may have been due to the fact that the experiment was carried out during extremely dry years, and emphasizes the importance of the particular hydrological regime in this general process. The trees sequestered large amounts of nitrogen in a two year period (330 kg N ha^{-1}).

A more extensive program of agricultural engineering has been studied by Mander (1991; Mander *et al.* 1989). This approach involves conceptualization of a watershed and sub-basins from a specifically landscape ecological basis. He found that 10 m wide strips of *Alnus* and *Salix* can absorb most phosphorus and about one-half of the nitrogen, lead and cadmium moving from farm fields toward a stream. He also found that 'vegetated bioplateaus' such as small islands in rivers and ponds can also filter and retain nutrients.

Transformations

Others have focused on the transformations of nutrients in the riparian zone. Brinson *et al.* (1980, 1983) reported on denitrification and phosphorous removal from over-bank flow in floodplain sediments.

Rhodes *et al.* (1985) quantified the uptake of nitrate in Washington. Pinay and Labroue (1986) documented denitrification in groundwater under riparian forest in France, and Pinay *et al.* (1989) found that differences in microtopography and soil moisture, as discussed above in relation to plant community composition, also affected the process of nitrification and denitrification. Fustec *et al.* (1991) documented the role of an abandoned meander channel, also discussed above in relation to internal structure, in denitrification of alluvial groundwater. Bartel-Ortiz and David (1988) used field and experimental techniques to examine the transformation of sulfur compounds. They found that flooding conditions affected the sulfur cycle because it leads to deposition of organic matter (which will be discussed below) and increased reduction.

A key point will be the change from aerobic to anaerobic conditions, how this changes in time and space, and the functioning of the requisite bacteria, including those that can change their functioning, in these two conditions. Pinay and Decamps (1988) summarized much of these concerns into a conceptual model, coupled with a field study, which noted the importance of the hydroperiod and also the sources of carbon as well as nitrogen. The capillary fringe is where the denitrification process is likely to be most important, and where the appropriate bacteria and the necessary carbon are likely to coincide. Ambus and Lowrance (1991) examined denitrification in riparian soils and found responses to added NO_3, especially NO_3 with carbon, but not to carbon alone. Surface soils were more active than subsurface soils.

Hill (1983, 1986; Hill and Warwick 1987; Hill and Shacklet 1989) has particularly concentrated on nitrogen transformations. His work is of particular interest because while the working of a simple linear filter along a river seems clear, he has put it in the context of the surrounding hydrological landscape, and the landscape-level processes of groundwater and surface water flows. Differences in the spatial relations of inputs and outputs and transport processes, rates, and directions affect the nutrient transformations, and thus their pathways and ultimate fates. These relations in the hydrological balance of riparian areas need further study.

A group of the key researchers in this field met at the quadrennial congress of the International Association of Landscape Ecology in July 1991 (e.g. Correll, Gilliam, Hill and Mander). A general conclusion is that nitrogen is effectively removed from the system while phosphorus is not (Correll *et al.* 1991). In terms of the spiraling concept, it would seem that again an alternative pathway is required in the model, i.e. denitrification. Basically, the spiraling concept was developed for a fixed amount of nutrient moving into and out of a reach of stream in the stream flow. The consideration of alternative inputs and outputs will embed the spiraling concept in more traditional systems models. Model development in this direction is also required by the importance of the local hydrological regime.

Nutrient effects

A question that arises is: what effect do these nutrient inputs have on the establishment, growth, and relative abundance of species in the riparian area? Riparian areas, especially in agricultural areas such as the midwestern USA, seem to be already rich in nutrients because they are depositional sites (e.g. Denaeyer-De Smet 1970). But the difference made by agriculture during the past 50 years is unknown. Mollitor *et al.* (1980) documented the spatial variability of nutrients in floodplain soils. They found differences between areas of vertical and lateral accretion. Peterson and Rolfe (1982*b*) documented the abundance of nutrients in floodplain soils relative to uplands. They noted important seasonal variations in floodplain nutrients which were significantly affected by flooding. Flooding leads to reducing conditions and an increase in pH, which reduces available phosphorus. Potassium was affected directly by leaching. But Peterson and Rolfe (1982*b*) noted that the effects on tree growth of the inputs of nutrients would be difficult to asses because of variation in other factors such as climate. Wittwer *et al.* (1980) showed, however, that inputs of fertilizer could have a significant effect on the growth of *Platanus occidentalis*, but the response on upland sites was much greater than on floodplains. Even this floodplain site, however, is one that probably received agricultural inputs in the past decades. Thus it may be that the long-term effects of such inputs have been significant.

Weiss *et al.* (1991) have found that over-bank flooding increases the available phosphorus in riparian sites, but that the ultimate fate depends on the soil type and on the specific responses of plant species, some of which store phosphorus in wood while others cycle it more quickly through leaf-fall. Pinay *et al.* (1992) noted that the processes of nitrogen and phosphorus dynamics depend on very specific hydrological and chemical conditions, which vary spatially and temporally. These dynamics affect their availability for plant growth and the results of specific measurement efforts. The differential response of species, however, is unknown and this information would be necessary to project the effects at the community level. In many sites in the midwestern USA, the understory of floodplain forests is dominated by *Urtica dioica*, a species well known to thrive on sites of anthropogenically increased nitrogen and phosphorus concentrations. In Iowa riparian forests, *Urtica dioica* forms an extensive, continuous subcanopy at a height of about 1 m (Figure 4.8*a*).

More general considerations of the processes of mass cascades of nutrients and carbon are typified by studies of drainage basins as ecosystems. The best example of this approach is the work undertaken at Hubbard Brook, New Hampshire (Bormann and Likens 1979). These studies examined the relations among precipitation, runoff, stream chemistry, and the dynamics of the vegetation in the basin. They directed their attention toward mechanisms by which an ecosystem regulates its export of matter, but they did not emphasize specifically spatial riparian functions.

Geomorphology

Some questions related to water quality are also related to geomorphology and geomorphological processes. While riparian zones may act as filters and trap sediment, they also may serve as a source area for sediment. Three areas of research should be considered: the sediment budgets of entire basins, the details of erosion and deposition along stream channels, and the role of riparian vegetation. The temporal dynamics of these forces are also important. From a geomorphological perspective, sediment spiraling would be considered at longer time scales than would be the case for ecological studies.

Basin sediment yield

One viewpoint to consider is the overall sediment budget of a large watershed. If we try to account for the sediment yield of the river basin at any point in time, two sources are considered: upland slopes, especially in agricultural areas, and river channels and banks. Once entrained, eroded sediment from upland slopes can quickly enter streams and be carried out of a basin, but as the studies above have documented, much of the sediment can be deposited in riparian areas where it is temporarily stored. During over-bank flooding, sediment is also deposited on these riparian areas. Even in plowed agricultural fields in the riparian zone, over-bank flooding does not erode additional material because the low slopes do not translate into enough tractive force to entrain new material, and these waters are often near capacity (the amount of sediment a river can transport at a given discharge) in any case (Trimble and Knox 1984). The other primary source for sediment yield, which may be a remobilization of the sediment trapped by the riparian filter, is channel and bank erosion. In many places this source may be more important than upland slopes. For example, Simon (1989) reported increased contribution of sediment from channel and bank erosion in channelized or straightened streams. He identified a series of stages during which different inputs and storages become dominant in such a system. The riparian ecosystem plays a role in this part of the sediment budget also.

As an example of the study of basin sediment yield, the upper Mississippi River valley has some interesting features studied by geomorphologists working on the problem in southwestern Wisconsin. Knox (1977) reported historical changes in stream form and sedimentation on floodplains in this area. He reported alluviation on the floodplain surface on the order of 0.5 to 4 m, contrary to the minor amounts of over-bank accretion postulated to be general by Wolman and Leopold (1957). He later was able to quantify actual deposition rates, and related these to changes in agricultural land use and especially to related channel adjustments (Knox 1987). He described a system in which agriculture led to large floods, especially in headwater and tributary areas. These larger floods, being more competent (able to transport material of large size), carried a coarser bed load; the coarse bed load and flooding together led to channel widening in these headwater and tributary streams. High rates of sediment deposition on the floodplains of these tributaries, initiated in the early period of agriculture, ended after the channels had adjusted to the point where they could contain the higher flows. In the lower valleys along the main trunk streams entering the Mississippi River, the story was different. These streams never had the competence to carry the coarse bed load, and so the armoring it would provide was absent and these streams cut downward to be narrow and deep relative to pre-settlement conditions. Over-bank flooding and sediment deposition on the floodplain continued in these reaches, however. Since the 1930s, sediment deposition has decreased owing to better agricultural conservation practices.

Trimble (1981, 1983; Trimble and Lund 1982) has emphasized the storage aspect of the overall sediment budget for a basin in southwestern Wisconsin. Trimble and Lund (1982) estimated the rates of upland erosion using the Universal Soil Loss Equation (Wischmeier and Smith

1965) and measured sediment deposition in reservoirs and on the floodplains in the basin. He reported that 50% of the human-induced eroded sediment had left the basin and that 50% had been deposited on floodplains. Through the use of floodplain cross sections and cores, he recorded different rates of sedimentation in historical times. He presented a model of the sediment budget for the basin with sources and sinks identified for the periods 1853–1938 and 1938–75. This model shows that although inputs from uplands have decreased by about 25% in the latter period, basin sediment yield remains the same because there is less deposition in the lower floodplains and because of contributions to the stream from the sediment stored in the upper main valley during the earlier period.

Others have also examined this problem. Wilkin and Hebel (1982) reported cesium-137 counts to measure a sediment budget for a basin in Illinois. Their conclusion was that much of the sediment eroded from uplands never leaves the basin, and instead is trapped and stored in depressions and along fences, roads and grass strips. They believe, however, that cropped floodplain surfaces are important sources for sediment yield from the basin. They measured annual losses of 40–60 tons per acre in one site and annual rates of 15–30 tons per acre commonly. This study was decried as unrealistic by Trimble and Knox (1984). The latter authors calculated that these rates would lead to losses of soil of 35 cm per 100 yr, and noted that in all of their observations in Iowa, Illinois, Minnesota and Wisconsin they found deposition, not erosion, on the floodplains. They noted that they had observed local erosion on floodplains that had been recently plowed, but concluded that these were isolated incidents because streams would not normally gain the tractive force necessary to entrain sediment in these areas because of the low gradient. Wilkin and Hebel (1982) noted, however, that in a forested floodplain they did record deposition, not erosion. The solution to this dilemma may lie in one or more of several areas. First, cesium-137 counts have been used in a variety of studies to document rates of soil erosion and deposition, but a number of assumptions, which may not be equally supportable in all areas, are required. The fact that Wilkin and Hebel (1982) did find deposition in the forested floodplain in contrast to erosion on the cropped floodplain indicates that the direction of these tests may have been correct. Trimble and Knox (1984, p. 1318) pointed to another fault of the Wilkin and Hebel (1982) study, that they did not specify very well the landscape in which they were working. They noted the failure 'to specify slope, relief, and other landform dimensions at their study site'. And they go on to comment that: 'the susceptibility of a valley floor surface to erosion, sedimentation, or stability at a point in time is much influenced by its relative geographic location within the hierarchy of the drainage network'. I would add that other factors of the landscape, such as land use and vegetative cover, need to be included in considering relative geographic location.

The determination of the inputs of sediment to rivers is important for the assessment of off-farm impacts of soil erosion and the best means to reduce both on-farm and off-farm impacts. Agricultural non-point source pollution has a major impact on the stream ecosystems in agricultural areas worldwide. The way in which these impacts accumulate has been detailed by Clark *et al.* (1985), but they did not consider impact on natural riparian areas. They did mention that aggrading rivers can create swamps and bayous, but saw any reduction in drainage of riparian areas as a detriment. They also mentioned the impact of overabundant sediment deposition on floodplain crops, but made no mention of sedimentation filling wetlands or similar areas.

Details of erosion and deposition

In relation to riparian environments, the two major concerns of fluvial geomorphology, erosion and deposition, are time- and place-dependent. In some mountainous environments erosion is the predominant factor, and deposition dominates in deltas. Between these two extremes, both erosion and deposition occur at any one place and the nature of the dynamic is still a question in geomorphology. The question of deposition versus erosion in the riparian zone may be the central issue of sediment yield. The spatial distribution of the processes includes an upper course, with rapid erosion of stream bed and banks, a middle course with spatially contrasting bank and bed erosion and deposition of sands and gravels, and a lower course where bank erosion occurs but silts and muds are deposited on the channel bottom. Except in deltas at subaqueous baseline, erosion will exceed deposition in all sections of the river, over long times, and the rate will depend on the difference from the base level.

Channels erode in three ways: headcutting, downcutting, and lateral erosion (Morisawa 1985). The latter two are of primary interest here. Downcutting is especially controlled by base level, and depends in part on the composition of the material. Downcutting will tend to occur in well consolidated materials. Notable instances of downcutting are evident in deep canyons cut into bedrock. Lateral erosion occurs in poorly consolidated material or when the stream bed is armored against erosion relative to the banks. In terms of the relation of geomorphology and riparian landscape ecology, lateral erosion in the middle reaches of river systems is of most interest because here the question of the riparian areas as a source or sink of sediment arises. The nature of the geomorphological process in this area is made up of the migration of meanders, with erosion on the cutbanks and deposition on point bars.

Migration of meanders and channel stability A more complex view of meander processes than is found in texts has been presented by Hickin (1974), among others. Meanders migrate downstream: erosion occurs on the outside of the bend, or cutbank, and is greater in the downstream portion of the bend. Deposition occurs at the inside of the bend on point bars: stream velocity is least here and hydraulics produce eddies and depositional sites. At any point in time the inputs and outputs of sediment to the stream may be in balance, but in most instances the movement of sediment into and out of riparian storage in these processes occurs in pulses. Meander forms are often compound and asymmetrical. Erosion at cutbanks does not occur gradually. More often the process is similar to erosion in gullies, i.e. slab and rotational failures (Murphey and Grissinger 1985; Springer *et al.* 1985; Okagbue and Abam 1986; cf. Bradford and Piest 1977). During periods of high water, the sediment is saturated and high pore water pressures occur. When the river recedes, the stream pressure that may have been holding up the bank is removed. High pressure in the pores is greater than the cohesive forces in the sediment, and bank failure occurs (Figure 5.5). Groeneveld and Griepentrog (1985) found a relation between groundwater, riparian vegetation and stream bank failure. This slump or rotational failure may not enter the stream immediately, but provides unconsolidated sediment that can more easily be entrained by higher water in the future. The erosion of banks may be as significant as upland erosion from agricultural fields in adding sediment and associated nutrients, especially phosphorus. Odgaard (1987) argued that 50% of the sediment leaving Iowa in rivers came from cutbank erosion (in contrast Oulman and Lohnes (1985) reported that most of the sediment yield from three basins in central Iowa was from sheet erosion, but this study was flawed), and Duijsings (1987) reported 53% for Luxembourg. The processes in riparian areas and the interaction with the river are affected by the changing shape of the stream channel. Knox (1987) reported such a change in southwestern

Figure 5.5 Without the support of vegetation, the bank of a river collapses following flooding, and the loose sediment can be entrained by a subsequent flood; White Breast River, Iowa.

Wisconsin, with channel widening in headwater streams and channel deepening in larger streams.

Depositional conditions In studies related to water quality, it was noted that sediment was deposited in riparian areas during over-bank flooding. This deposition leads to vertical accretion in addition to the sediment deposited in point bars. Such deposition, however, is not spatially homogeneous on the floodplain. Alexander and Prior (1971) documented differences in the sediment texture deposited at natural levees, and on the swales and ridges of the floodplain interior. Rates of aggradation on levees were near 10 mm yr^{-1} over *c*. 800 yr. In a swale, rates were 0.27 mm yr^{-1} between 7000 and 6000 BP; 6.0 mm yr^{-1} between 3000 and 2000 BP, and 1.9 mm yr^{-1} between 2250 and 1250 BP. On ridge sites, values were 0.26 mm yr^{-1} from 10 100 BP to present 3 km from the river; and at a site 0.4 km from the river 0.31 mm yr^{-1} between 5000 and 2500 BP and 0.18 mm yr^{-1} since 2500 BP. Phillips (1989*a*) noted the geomorphic function of riverine wetlands that trap and store sediment. Using data from seven basins from across the USA, two in Europe, and one in Australia, he calculated that in large basins 15% of total upland eroded sediment and 50% of that which reached streams became trapped and stored in downstream wetland areas.

The transfer of suspended sediment from the main stream to the floodplain has been considered in terms of a diffusion–dispersion model (James 1985). The primary factor affecting this sediment distribution is flow velocity. It is well known that when a river overflows its banks, velocity drops at the edge of the floodplain, and this leads to the deposition of levee sediments. Over the floodplain as a whole, however, the situation is more complex. Abandoned channels, levees, swales, and ridges will have different flow conditions and may have different rates of deposition. Moore and Burch (1986) considered the kind of complex topography that would affect sediment deposition by over-bank flooding. James (1987) modeled sediment deposition in common channel systems. He demonstrated the importance of turbulence and described the vertical and transverse distribution of sediment in over-bank flood flows. The existence of turbulence means that quantitative prediction may not be feasible (Malanson *et al.* 1990). Several studies in the midwestern USA have documented rates of sedimentation in wetlands and ponds such as abandoned channels on the floodplain. Bhowmik and others (Bhowmik and Demissie 1986, 1989; Demissie and Bhowmik 1987; Bhowmik *et al.* 1988) documented such sedimentation in Illinois. They reported losses in the volume of such areas ranging from 30 to 100% during the period of agricultural development in historical times. While these studies are directed toward sediment erosion and yield problems, they reveal a significant geomorphic effect in riparian areas which should have notable ecological consequences for these riparian elements (Eckblad *et al.* 1977). Cooper and McHenry (1989) and Hupp and Morris (1990) found similar results further south in Mississippi and Arkansas, with highest rates of sedimentation occurring in sloughs, using dendrogeomorphological techniques similar to those pioneered by Sigafoos (1964). Sparks *et al.* (1990) reported accumulation of fine sediment in abandoned channels and backwaters along the Illinois river.

Role of vegetation

One of the notable features of sediment erosion and deposition, and the effect of flow on them, is that they must be affected by vegetation. Fitzpatrick and Fitzpatrick (1902) noted that the riparian environment plays an important role in affecting this dynamic, and Hickin (1984) identified five ways in which vegetation affects fluvial geomorphology: flow resistance, bank strength, bar sedimentation, formation of log-jams, and concave bank-bench deposition. The relations among riparian vegetation, deposition and erosion, and flooding must be emphasized, but the processes by which these interact has rarely been considered. Nanson (1980) and Jackson (1981) described detailed sedimentological sequences associated with floodplain formation but neither presented a role for floodplain vegetation other than to contribute organic matter. Nanson and Beach (1977) discussed the interaction of sediment and vegetation in floodplain formation and succession in a subalpine area in Canada. They found that rates of sedimentation affected the establishment of tree seedlings, and that the trunks of trees probably affected the sedimentation. They hypothesized that deposition decreased when more open stands developed and created less friction for over-bank flows.

In previously cited studies of the invasion and colonization of point bars by pioneer species, authors observed that the vegetation helps to extend point bars (e.g. Craig 1992). Throughout a floodplain forest sediment deposits can be found following extensive flooding. Most often they are fine-grained, but sand lenses can be deposited in this way near the channel. Shelford (1954) felt that vegetation played a major role in controlling the rate of aggradation and thus its own topographic position in extensive forests on the Mississippi River.

In the Southwest Robinson (1958), Hadley (1961), and Graf (1985) have noted that invading phreatophytes cause sedimentation, which leads to a wider spread of over-bank floods, which in turn favors the expansion of the phreatophyte vegetation. This dynamic interaction over a short time span is probably indicative of processes operating over a longer period on major alluvial rivers. Graf (1980, 1982, 1985) has discussed the importance of riparian vegetation in maintaining horizontal stability of river channels in the southwestern USA and the response of riparian communities to changes in flood regime caused by dams. He reported major channel instabilities in the Southwest that are qualitatively different from those of a major alluvial river in that a floodplain *per se* may not exist. Conversely, because of downcutting, erosion and refilling, a floodplain may disappear and later be recreated. A set of complex interactions was examined by Groeneveld and Griepentrog (1985) in California. They linked the depletion of groundwater to the loss of riparian vegetation and thus to erosion in a causal sequence.

Trent and Brown (1984) reviewed river stability considerations and identified vegetated banks as one of the few factors contributing to stability. Riparian vegetation is universally considered to decrease the rates of bank erosion (Wolman 1959), except where woody debris may lead to scouring (Duijsings 1987; Robison and Beschta 1990*a*). Parsons (1963) identified three means by which vegetation protects a streambank from erosion: by reducing the tractive force of water, by protecting the bank from direct impacts, and by inducing deposition. Smith (1976) found that roots of meadow grasses and willow scrub produced low rates of erosion, which he contrasted to worldwide rates ranging up to 750 m yr^{-1}. Smith (1976) and Bray (1987) have also shown how vegetation can contribute to channel stability by reducing rates of bank erosion. Odgaard (1987), while concentrating on less vegetated banks, noted that cutbank erosion rates along forested reaches were much lower, and Ikeda and Izumi (1990) provided a theoretical analysis of bank stabilization by vegetation. The roots of the vegetation provide a mechanical support and reduce the instance of shear failure. The nature of rotational failures in streambanks has been observed to vary with the depth of rooting (Okagbue and Abam 1986). Observations of river banks in Iowa indicate that where banks are being eroded and trees are falling into the river a complex pattern of protection of the bank at the point of fall by the tree alternates in space with intensified erosion just downstream, and in time with renewed erosion when the tree is finally washed away. Hupp (1987; Hupp and Simon 1986; Simon and Hupp 1986) reported dendrochronological studies of the effects of channel straightening in loess soils in western Tennessee. These studies document rates of channel widening and downstream sedimentation. As noted above, the erosional processes identified are similar to those of gully erosion, i.e. slab and rotational failures. Butler *et al.* (1992) have hypothesized that the increase in beaver in midwestern streams will alter the relation of vegetation and cutbank stability by decreasing large tree-falls while good root networks remain little changed.

A need exists for a better developed theory focusing on sediment delivery ratios (e.g. Duijsings 1987). By increasing the roughness and slowing the velocity of over-bank flows, riparian vegetation promotes the deposition of sediment (Li and Shen 1973). As in the studies on water flow over vegetated floodplains, the true complexity needs to be considered. The actual effects of individual species, their sizes and physiognomy, and their distribution on boundary shear stress, flow velocity and surface roughness need to be identified (cf. Arcement and Schneider 1989). These are the factors that affect the spiraling at the functional level.

Role of woody debris

Harmon *et al.* (1986) recently reviewed the presence and functions of woody debris in temperate regions. Much of the review was devoted to debris in lotic systems. They identified three areas in which woody debris interacted with geomorphology: in stream morphology, in the storage of matter, and in the storage vs. movement of the debris itself. There seem to be two distinct scales at which woody debris operates. In small streams, particularly in mountains, trunks can block a channel quickly and form a dam or 'log step'. These small-scale dams can trap large amounts of sediment. Several studies demonstrate this relation. Hack and Goodlett (1960) described how trees helped form debris dams that resulted in the incisement of channels in Appalachian mountain streams. Heede (1972) found that log steps in Colorado mountain streams resulted in the addition of gravel bars as the streams tried to adjust their gradient. When the log steps were removed, the gravel bars grew as bed load movement increased (Heede 1985). Keller and Tally (1979) and Mosley (1981) reported on their effective trapping of bed load. Marston (1982) and Duijsings (1987) noted that they dissipate energy and collect and store sediment. In addition to log steps, Kochel *et al.* (1987) showed that log dams parallel to the channel can trap sediment and build what appear to be terraces.

Other experiments in which debris has been removed from streams show that large amounts or sediment are then mobilized without a return to a new equilibrium, at least over the course of the study (Beschta 1979; Bilby 1984), and that when such dams break the consequent downstream erosion is important (Gurnell and Gregory 1984; Gregory *et al.* 1985; cf. Butler 1989). Andrus *et al.* (1988) reported that debris contributed to pool formation, and in Alaska, Robison and Beschta (1990*a*) found an association between woody debris and deeper pools, and they attributed the heterogeneity in the stream to woody debris, in part. Beaver dams also capture and store sediment (Allred 1980; Sinitsyn and Rusanov 1990). Early removal of beaver probably led to increased sediment yield (Parker *et al.* 1985), and the reintroduction of beaver into some areas has led to increased sediment storage (Brayton 1984; Apple 1985). This process has in effect been a reversal of the debris removal experiments. Smith and Shields (1990), comparing in-channel conditions between reaches cleared of debris and controls, found that debris significantly increased habitat diversity.

In larger rivers the woody debris may not form complete dams, but the entire form of a river system can be altered by this process (Keller and

Swanson 1979). At a smaller end of this scale, Keller and Melhorn (1973) cited a log dam as responsible for the cutoff of a meander on a stream in Indiana. Harmon *et al.* (1986), in describing the spatial patterns of woody debris in lotic systems, noted that debris in mountain regions tended to be strongly clumped and associated with particular points of geomorphological deposition such as the inside of meander bends at the heads of midstream bars. This process raises a question of mutual causality. The information on vegetative succession, alluvial deposition, and the role of woody debris remains unconnected. Hypothetically, woody debris promotes deposition, soil development, and succession to forest, which in turn promotes further deposition and succession and then provides a ready source for woody debris in the future. Malanson and Butler (1990) documented the interaction of woody debris and the storage of coarse gravels, fine sediment, and organic matter in longitudinal channel bars. Although there seems to be an interaction, we could not clearly separate cause and effect.

At the larger end of the scale of large rivers, Sedell and Froggatt (1984) described the change in channel pattern in the Willamette River, OR over a 100+ yr period during which woody debris was removed. Without the presence of large logs blocking and diverting flow, the channel became more incised and a complex pattern of channel and sloughs was reduced to a single channel. Sedell and Froggatt (1984) noted that the older system presented a modification of the River Continuum Concept wherein the multiple channel system was more characteristic of lower-order streams than would be the case today. They quoted historical sources that reported on the inconstancy of such a channel system. Triska (1984) reported on the existence of a large log-jam on the Red River in Louisiana that covered a reach of up to 225 km. This condition had probably been common since the fifteenth century. The raft, as it was called, resulted in extensive geomorphological features, notable bayous, sloughs, and lakes, that were reduced or eliminated when the raft was finally removed in 1873, after decades of attempts. As Triska (1984) noted, it is no longer possible to study the role of such debris in large rivers because they have been cleared for commerce. The process by which woody debris gets into streams is under study. Robison and Beschta (1990*b*), Van Sickle and Gregory (1990) and Malanson and Kupfer (1993) have provided some mathematical tools for identifying trees that might contribute directly to a stream and for estimating probabilities of inputs of woody debris.

Equilibrium or nonequilibrium geomorphology

The equilibrium paradigm in fluvial geomorphology has been abandoned. Efforts now address the lack of equilibrium between geomorphological processes and climatic change and ecological change. The most notable cases of disequilibrium occur in what are called paraglacial conditions, where sediment budgets are related to past glaciations (Church and Ryder 1972). Another case occurs where major and extensive changes in land use over a short interval, particularly in the twentieth century, have created nonequilibrium conditions (Trimble 1983). Short-term natural fluctuations, or pulses in fluvial processes, may also be interpreted as nonequilibrium or dynamic equilibrium conditions (e.g. Nanson and Hickin 1983).

Paraglacial geomorphology Church and Ryder (1972) identified major terrace deposits in the Rocky Mountains as instances of nonequilibrium geomorphology, which they termed a paraglacial condition. During the period of Pleistocene glaciation, the production of sediment through weathering processes greatly exceeded its transport out of the basins. In the Holocene, this source of sediment has continued to affect the sediment yield of the basins, and sediment transport now exceeds the rate of weathering. Church and Slaymaker (1989) and Church *et al.* (1989) have documented increases in sediment concentration in rivers as one moves downstream in British Columbia, similar to results found by Trimble (1983) (discussed above and below). This increase contradicts the notion that sediment is deposited in intermediate locations, and instead indicates that at this time those intermediate storage areas are acting as sources. Church identifies these as secondary remobilization of Quaternary sediments. Malanson and Butler (1990) examined one aspect of this nonequilibrium condition in Montana. We found that major pulses of gravel in the river are still responding to the massive amount of sediment stored in terraces (Figures 2.1 and 3.9). We did not attempt to document specific source areas for gravel pulses, but exposed cutbanks on the order of 100 m high are common. Paraglacial conditions may also be said to exist in areas of continental glaciation. The form and size of river basins has been related to Pleistocene and early Holocene fluvial conditions, and the phenomenon of the underfit river is common (e.g. Salisbury *et al.* 1968). The sedimentological conditions left by previous glaciations, and whether these are best considered as equilibrium or nonequilibrium phenomena, however, have not been considered. In

Iowa, most of the landscape is made up of thick deposits of Illinoisan till and loess and, in one area, Wisconsinan till. These deposits certainly affect the sediment yield of the basins, but the concept of paraglacial conditions, if it is to mean anything, given that all sediment regimes of basins are conditioned by the past, may be applied best in mountains.

Historical conditions Bravard (1987) presented a detailed description and analysis of the human impacts on the Rhone between Geneva and Lyon. It is clear from this and associated studies (e.g. Bravard 1982, 1983; Poinsart *et al.* 1989) that while the river has adjusted to the glacial conditions much more than have subalpine rivers, the influence of human occupation for several centuries has had a lasting effect. It is difficult to point to the disequilibrium between a particular impact and a particular response because the impacts are widespread in space and continual in time. Impacts have included gravel mining, channelization, dams, dikes, and diversions. Responses have included channel shifts and abandonment and an overall simplification of the spatial diversity of the river and the riparian area. The sediment regimes and the instances of cutbank erosion and point bar deposition have been greatly affected by these human impacts.

Twentieth century conditions The sediment budgets of the upper Mississippi River drainages are still in response to changing land use. In the nineteenth century agriculture was introduced to the prairies and woodlands of this region. As a result extensive erosion occurred. Much of this material eroded from uplands was deposited in riparian areas. Since the 1930s, improvements in farm management have reduced the inputs of sediment from upland fields. A question that arises is whether the runoff now has the capacity to carry more sediment, which will be entrained from its storage in riparian areas. This question is important for the assessment of the effects of in-field and off-farm soil management practices on water quality. If sediment eroded decades earlier will replace sediment that would have eroded from farm fields, or is now trapped in riparian filters, then attention to the stabilization of riparian storage will be needed. Rhoads and Miller (1990) illustrated a sediment budget methodology that may be useful for accounting for the effects of sediment management on the geomorphological stability of river channels and adjacent floodplains. They found that the initial stages of a project to construct new wetlands along a river in Illinois did not have an effect on downstream channel stability, but several more years of monitoring will be needed to assess the effects.

The inclusion of temporal variation in the study of riparian landscapes is important. Nonequilibrium effects imposed by human activities are probably widespread. Phillips (1988) demonstrated the effect of including change in hydrological simulation modeling. These cases of nonequilibrium fluvial geomorphology affect the distribution of landforms and directly and indirectly the ecological processes and patterns on a landscape. Nonequilibrium relations in geomorphology may lead to unpredictable long-term landscape development (Malanson *et al.* 1992). The application of the spiraling concept to geomorphological systems with temporal responses or time-lags of millennia may add a useful approach to this nonequilibrium study, but may be stretching an analogy too far.

Cascades of carbon

While photosynthesis is the origin for organic carbon in the riparian ecosystem, this process is not distinctly riparian in any way. The movement of organic carbon fixed in plant and animal tissue does, however, have some specifically riparian characteristics. The most notable are the production and transport of large woody debris and leaf litter. The former has been addressed as a geomorphological agent, and in part could have been discussed in the section above, but both have been considered as carbon inputs to in-stream ecosystems. In fact, the carbon cascade of riparian areas has been studied primarily in its role as an input to the aquatic realm; for example, Mulholland (1981) noted that the floodplain could be the primary area of carbon storage for and the primary source of carbon input to lotic ecosystems.

Productivity

Actual gross or net primary productivity is rarely measured in riparian environments; although biomass is more commonly reported, it is usually above-ground plant biomass only. In his wide-ranging review of riparian forests, Brinson (1990) reported above-ground biomass from 17 freshwater sites, of which only 9 also provided data on below-ground biomass. While in forested riparian areas the non-plant biomass may be an insignificant proportion, the below-ground or root biomass may be important, especially in phreatophyte systems such as found in the southwestern USA where tap roots may be very deep (Gatewood *et al.* 1950). Most of the reports on productivity in riparian forests have come from eastern North America, but the most comprehensive report is from Czechoslovakia.

Vyskot (1976*a*,*b*) reported below-ground biomass, both leaf and woody above-ground biomass, litterfall, wood production, and total biomass production of a riparian *Quercus robur* forest in Czechoslovakia. Vyskot (1985) reported an expansion of this same study. Fifteen individuals (7 *Q. robur*, 6 *Fraxinus* spp., 2 *Tilia cordata*) were analyzed in detail. Biomass was measured for stems, branches, shoots, roots, leaves and bark. Above- and below-ground vertical distribution, diameter distribution and canopy distribution of parts were recorded, with a total biomass of 36 kg m^{-2}, of which 4.6 kg was below ground, and 31.4 kg was above ground. Vasicek (1985) reported biomass values for litterfall and for the shrub and herb layers of this forest. Because of differences in shrub density, this biomass ranged from 39 to 98 kg m^{-2}. Zejda (1985) examined small mammal secondary productivity, Zajonc (1985) addressed earthworms, and Grunda (1985) studied decomposition. Lesak (1985) supplemented the study with productivity measurements of a riparian meadow in the same area. This study, headed by Emil Klimo, provided the best basic information on ecosystem structure of a complete riparian forest area. Although a second volume of this research has recently been published (e.g. Palat 1991; Prax 1991*a*,*b*; Vasicek 1991) the infrastructure of the study fell into disrepair in the 1980s; the team intended to revive the work, but a complete renewal of the research is hampered by a lack of funding (E. Klimo, personal communication). The generality of these results, and their integration into an overall system with well-defined forcing functions, depends on this renewed study.

Brinson (1990) provided an extensive review of the literature on biomass and production in riparian forests. He reported a range of biomass of 25.5 to 608 t ha^{-1}, with most reports between 100 and 300 t ha^{-1}. Above-ground net primary productivity ranged from 6.48 to 21.36 t ha^{-1} yr^{-1}. For those studies reporting, litterfall amounted to 47% of the total. These litterfall levels are quite high. Brinson (1990) noted that although a number of studies exist, they are still too few to draw conclusions about general patterns because of the complex interaction of hypothetical controls on production, i.e. hydroperiod and soil moisture, nutrients, redox status, and mechanical damage. It can be concluded, however, that the productivity of these particular floodplain forests is high relative to that of nearby upland sites.

Total biomass on riparian sites has been measured both by harvest and by estimations such as litterfall (Killingbeck and Wali 1978) and parabolic volume. Parabolic volume estimates depend on incremental growth in tree diameter. These measures may seem straightforward, and they have been used with some success (e.g. Taylor 1985), but many riparian species are much more difficult to analyze by dendrochronology than are the coniferous species around which the tree-ring methodology has been developed. Dendrometers, measuring circumference increments, may be more useful, but they provide only one year's datum per year, whereas the tree-rings can provide data well into the past. Mitsch *et al.* (1991), in a report based in large part on Taylor's (1985) dissertation, described differences in riparian forest productivity among three 'hydrologic landscapes' in Kentucky. While this is a useful ecosystem study, the 'landscapes' described are tesserae, not even landscape elements. To describe them as landscapes misinterprets the fundamental concepts of landscape ecology. That aside, they reported average net primary productivity of 1307 g m^{-2} yr^{-1} for seasonally flooded sites, 579 g m^{-2} yr^{-1} for sites with slowly flowing waters, and 205 g m^{-2} yr^{-1} for sites with stagnant waters.

From a very different perspective, Dunham (1990*a*,*b*) has examined the productivity of riparian woodlands along the Zambezi River in Zimbabwe. A major emphasis of his work was on the food supply for mammals in Mana Pools National Park. He documented the productivity and seasonality of herbaceous species in eight types of tesserae. The rate of consumption of biomass in grasslands was high, varying between 53 and 99% of annual production, and termites were the most important consumers. He also recorded that productivity of *Acacia albida* fruits depended on both precipitation and early feeding by baboons; the fruits were also eaten by buffalo, eland, elephant, kudu, and waterbuck (Dunham 1990*b*).

Woody debris

The ecological functions of streams and rivers are greatly affected by the input of organic matter from the surrounding vegetation. The geomorphological aspects have been discussed above. The effect of the drainage basin as a whole on stream dynamics has been cited repeatedly (Ross 1963; Likens and Bormann 1974; Hynes 1975). Lienkaemper and

Swanson (1987) described in some detail the introduction and movement of debris into mountain streams in Oregon. Windfall was found to be an important agent in these streams, as opposed to bank erosion. They did not report on any further effects on stream processes. The storage and movement of debris in streams and rivers has been only briefly described in other reports (Swanson and Lienkaemper 1978; Bryant 1980; Bilby 1984). Swanson *et al.* (1976) outlined the major ecological concerns regarding woody debris in mountain streams, and Swanson *et al.* (1982) described in detail the ways in which near- and in-stream debris could act to regulate functions between the aquatic and terrestrial zones. They described in more detail some of the effects on the ecological processes of woody debris in small mountain streams in Oregon. These streams were often narrower than the length of the debris. Detailed maps before and after debris movement provided a basis for assessing the contribution of organic matter to the stream ecosystem by this route as well as the effect of the debris as dams affecting invertebrate and vertebrate lotic communities. Bilby and Likens (1980) and Bilby (1981) reported an investigation of the role of debris dams in streams on the export of organic carbon and other material. Carlson *et al.* (1990) found similar amounts of debris and relations between debris and stream geomorphology in logged and undisturbed forests in the northwestern USA, and concluded that logging had had little impact. Beaver also provide large inputs of debris to streams.

Litter

Bell and Sipp (1975) and Bell *et al.* (1978) measured litter composition and processing along an elevation gradient in a riparian forest in Illinois. They concluded that the higher litter amounts found in the higher areas, which were never flooded, were due to three factors: greater input from on-site productivity, lower rates of decomposition on these drier sites, and less transport away from the site. This study emphasizes the three aspects of litter processing that need to be addressed on riparian sites: productivity, decomposition, and transport.

On-site production of litter

Only a limited number of studies of litter production on riparian sites have been completed, but the general conclusion, contrary to the studies of Bell and Sipp (1975) and Bell *et al.* (1978), is that riparian forests produce more litter than their upland counterparts. Some work has compared constantly flooded swamp forest with seasonally flooded riparian and upland forests, and the general conclusion is that the productivity of riparian, flowing-water forests exceeds that of standing-water flooded forests, which in turn exceeds that of uplands (Conner and Day 1976; Brown *et al.* 1979; Brown and Peterson 1983). Notable differences within the same reach of riparian forest are based on topographically related site conditions (Killingbeck and Wali 1978) (these gallery forests may include vegetation on valley side slopes which is not strictly riparian). This greater productivity and complexity means that the litter dynamics of riparian forests are some of the most interesting to study.

Adis *et al.* (1979) measured litterfall in central Amazonia. They compared biomass and elemental constituents of litter produced during both dry and inundation phases of igapo, and found that nutrient concentration in leaf litter was reduced during the flooding season. They also distinguished among leaves, woody material, and fruits in the litter. Leaves comprised 79% of the annual production, wood 14%, and fruits 7%. Two-thirds of the biomass in all three categories fell during the inundation season (April–August). The total litter production averaged 676 g m^{-2} yr^{-1}.

Peterson and Rolfe (1982*a*) documented monthly and more frequent litterfall patterns in riparian and upland sites in Illinois. They found high rates of litterfall in both sites (797 g m^{-2} on the floodplain and 809 g m^{-2} on the upland), but noted more variability on the floodplain site and the importance of specific events, such as an ice storm and a wind storm, on the fall of woody material. Shure and Gottschalk (1985) reported litterfall patterns for a floodplain forest in South Carolina. They distinguished among leaves, wood, and reproductive parts, and also among species. They sampled at bank edge, and at 30 and 60 m from the bank, and at an upland site, and they sampled in three successive years. Sites near the stream produced less leaf litter than sites far from the river, but wood, reproductive, and total litter were not significantly different (total litter range: 571.8 g m^{-2} yr^{-1} at the bank to 667.2 g m^{-2} yr^{-1} at 60 m). They determined that, as in some earlier studies, litter production in the riparian zone exceeded that of upland sites, and the productivity on the floodplain was greatest not in the wettest areas but in what was termed a transition area.

Killingbeck (1986) found that geographical variations in riparian litterfall were due in large part to precipitation and the length of the growing season. Muzika *et al.* (1987) compared the overall productivity and litter production of undisturbed and successional (15 yr old) riparian forest of both a small stream and a larger river in South Carolina. They reported no significant differences in the total litter production, but there may have been significant differences in the distribution of litter among components. While little difference appears between the successional and undisturbed stream sites, on the riverine sites the leaf and stem litter of the successional sites exceeded that of the undisturbed sites by 1.66 and 1.54 times, respectively, while the flower and fruit litter of the undisturbed site exceeded that of the successional site by 6.15 times.

Chauvet and Jean-Louis (1988) also found higher levels of floodplain litter deposition (4.7–6.0 t ha^{-1} yr^{-1}) than in geographically related uplands. They noted, however, that they, like other researchers, did not include the boles and stems of trees that would be added by tree-fall events. Gurtz *et al.* (1988) reported a longitudinal variation in litterfall inputs to a prairie stream. Because they were dealing with an intermittent stream with headwaters in grasslands and increasing woody vegetation to shrubs and trees as one proceeded downstream, they found an increase in litter production and floodplain storage in this direction.

Decomposition

The major contention is that litter decays faster on riparian or floodplain sites because of the increased moisture. Peterson and Rolfe (1982*a*) documented this difference between riparian and upland sites in Illinois. Irmler and Furch (1980) reported decomposition rates in both igapo and varzea of central Amazonia. They concluded that different elemental concentrations led to different effects of inundation on the rate of decomposition. The high rate of decay has been documented for in-stream experiments using riparian leaf litter (e.g. Mathews and Kowalczewski 1969; Tate and Gurtz 1986; Hill *et al.* 1988). Furch *et al.* (1989) performed such an experiment using tanks of water and concluded that the Amazonian riparian forest acted as a significant pump of nutrients from the sediments into the river. Factors other than moisture may also affect litter decomposition. Most notable would be the activity of invertebrates. Neiff and Poi de Neiff (1990) found a litter half life of 20 days over a 38 day study period at a riparian site in Argentina. Mayack *et al.* (1989) studied the effects of burial and floodplain retention on litter decomposition. They noted that in winter buried litter decomposed more slowly, but in spring invertebrates increased the rate for buried samples, so that the overall rate for surface and buried samples was similar. They concluded that retention of the litter on the floodplain over the winter and into spring did not greatly affect its rate of decay once transferred to the stream, but they thought that this storage meant that the organic matter became available to stream animals at a time when direct leaf fall would be minimal. Gurtz and Tate (1988) documented higher rates of decomposition with an increase of inundation frequency and duration. Shure *et al.* (1986) also found higher rates of decomposition on the floodplain than on an upland in South Carolina, and differences in the rates of leaching of certain nutrients. In addition to the simple effects of moisture, they concluded that species-specific differences in leaf composition also affected these rates.

Brinson (1990) summarized results from several studies, many of which were *Taxodium* stands (e.g. Duever *et al.* 1975; Burns 1978; Nessel 1978), and found little geographical variation or even variation in moisture conditions that explained rates of decay because of differences in the type of material. He computed the half life for litter loss for 32 cases and found the mean to be 1.19 yr. The median values (sixteenth and seventeenth in rank order) were 1.00 and 1.03 yr. Whether this exponential decay formula is appropriate may be questioned because the annual decay coefficient is probably not constant as size and structure of the litter changes. The results do indicate, however, that contrary to Brinson's (1990) interpretation of the same table, year to year accumulation may be significant.

McArthur and Marzolf (1987), in documenting the rates of release of nutrients into a stream from riparian litter, found differences among species in rates of decomposition, and cited the amount and timing of precipitation as well as the condition of the litter itself and the antecedent hydrologic conditions at any point in time. Chauvet (1988) followed the reduction of litter of *Salix alba* over 40 weeks in three aquatic (fast-running, slow-flowing, and standing waters) and two terrestrial (floodable and nonfloodable) situations. He found little difference in the pattern of reduction in mass, carbon, nitrogen, cellulose or lignin. He did note, however, the circumstances that might lead to similarities, e.g. oxygen vs. macroinvertebrates, between the aquatic and terrestrial areas, and he concluded that it is necessary to differentiate among landscape elements within a river corridor in order to examine the overall ecosystem function. E. Chauvet (personal communication) complicated this view, however, with findings that decomposition was higher in the aquatic and driest sites because of burial by sediment in the floodplain site. He found no differences among *Alnus*, *Populus* and *Salix* leaves. He now believes that hyphomycetes may be the single most important factor in rates of decomposition (cf. Chauvet and Merce 1988).

Nortcliff and Thornes (1988) related the decomposition rates of litter in central Amazonia. They found that stream hydrology played an important role. Soil moisture storage capability in the dry season and the presence of saturated soils affected the amounts of organic matter present. They concluded that variable source area saturation of the soil was important in their study area, but distinguished between its position in the system and that of lower-order areas, in which overland flow would be important, and higher-order areas where over-bank flooding would be the dominant factor.

Another, inorganic, pathway for carbon *per se* must be considered at this point. Wetlands emit methane to the atmosphere, and riparian wetlands and flooded forests contribute to this flux. Naiman *et al.* (1991) documented a flux of 8–11 g C m^{-2} for areas inundated by beaver activity in Minnesota.

Transport of litter

Most studies of litter production have not determined how much enters the river, so transport processes need elucidation. Studies of litter have, however, observed transport effects that were not directly a subject of research design. Adis *et al.* (1979) observed removal of litter from the floor of igapo in Brazil during the flood season. Peterson and Rolfe (1982*a*) also reported that litter on their sites was removed during spring floods, and this circumstance obviated the possibility of collecting forest floor litter prior to peak autumn litter fall.

Hardin and Wistendahl (1983) found significant spatial variation in litter accumulation in a floodplain forest in Ohio. They reported that accumulation was greatest next to trees, specifically on the upstream side. They attributed this pattern to the direct effects of flooding, and to their indirect effects which create depressions around trees through erosion where water has accelerated. In Iowa floodplain forests the redistribution of litter under the riparian forest is clear, and is similar to the pattern reported by Hardin and Wistendahl (1983). This pattern of redistribution and especially the collection of litter in specific cases is a microcosm of the effects of flooding on large woody debris (cf. Harmon *et al.* 1986; Malanson and Butler 1990).

Leaves may either fall directly from trees into streams, or, because they are relatively buoyant, be carried into streams during floods.

Brinson (1977; Brinson *et al.* 1980) reported on this pathway through a floodplain wooded swamp. Shure and Gottschalk (1985) reported data for the inflow and outflow of litter in over-bank flooding of a riparian forest. They did not find any clear pattern over three years of study. In one year inflow of litter exceeded outflow, and in the other two years outflow exceeded inflow, but total fluvial exchange was not significant (0.68% of the three year total litterfall), and over the three years nearly cancelled itself out (0.18%). High output of leaf litter during peak floods was balanced by input during lesser floods, and the long-term flood regime would need to be examined in order to assess the overall flux. The authors cited the actual season of flooding as an important factor. Dawson (1976) found that leaf litter was trapped by bank vegetation and did not directly enter the ecosystem of chalk streams in England, and Nilsson and Grelsson (1990) found that the movement of litter and its trapping within the riparian area, but reported that litter and woody debris were not correlated with riparian features such as topographic slope or stem density.

Although not fluvial in nature, another export of litter from the riparian zone is that of direct fall from trees into the river. Chauvet and Jean-Louis (1988) found that much of the litter from a stand of *Salix alba* fell directly into the river, and that while floods redistributed litter on the floodplain, they were not a major source of export from the floodplain to the river.

Grubaugh and Anderson (1989) distinguished among the components of coarse and fine particulate organic matter (CPOM and FPOM) and dissolved organic carbon (DOC) entering the Mississippi River. They found seasonal variation, with the riparian forest contributing DOC during autumn leaf fall and FPOM during spring flooding, but noted that as spring floods receded FPOM levels dropped owing to deposition in the floodplain forest. This deposited FPOM led to increased DOC later being returned to the river. McArthur and Marzolf (1986) reported longitudinal variation in the processing of DOC in a prairie stream depending on whether the leachate was from trees or grass. Fiebig *et al.* (1990) found that the soil water in the riparian zone was a direct source of DOC for the stream.

Modeling the carbon cascade

Malanson and Kupfer (1993) used a forest simulation model to project productivity, woody debris and leaf litter of a riparian cutbank edge site.

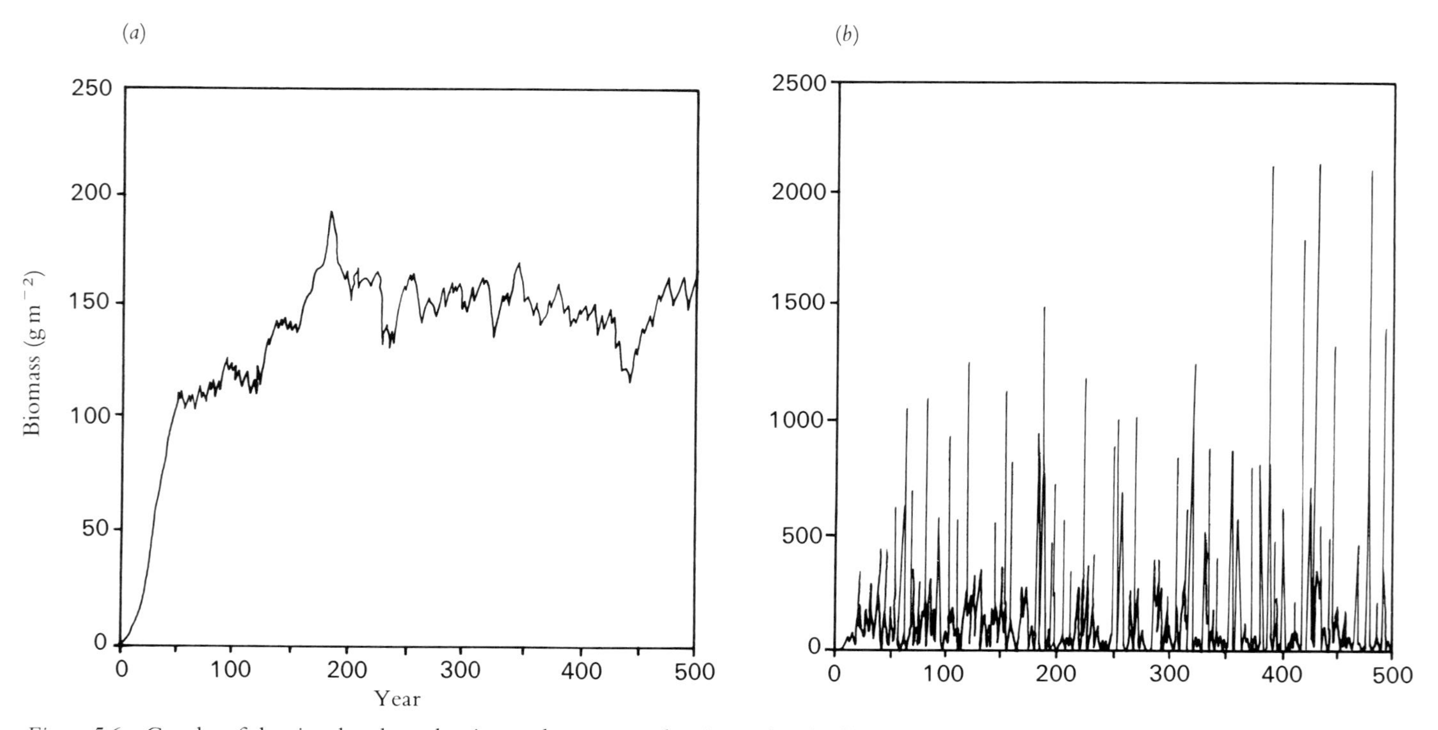

Figure 5.6 Graphs of the simulated production and export to the river of (*a*) leaf litter and (*b*) woody debris at a riparian cutbank edge; from Malanson and Kupfer (1993).

We altered the basic model (FORFLO) (Pearlstine *et al.* 1985) to examine differences between riparian interior sites and those of a cutbank edge where increased light leads to different species and productivity. We projected values of wood and leaf production, storage on site, decomposition, and export to the river (Figure 5.6). We examined projections with and without the specific edge effects on productivity, with high and low flood regimes, and with and without net export of woody debris and leaf litter during flood events. Our projected values for what would be a riparian forest interior site are similar to those reported by Mitsch *et al.* (1991), but the edge effect increased the amount of leaf litter and woody debris produced on the edge site by 22% and 44%, respectively. We combined several aspects of transport in the simulation model. We included leaf fall directly in the river from trees a distance less than their own height to the river, and projected that 32–33% of leaves produced on a 1/12 ha site follow this pathway. We used a probability function that trees growing on the outside would fall directly into the stream, and we projected that 39–42% of woody debris (the sum of all trees that die on the site) enters the stream in this way. The indirect effects of flooding were minor. At higher flood levels the amount of leaf litter and woody debris entering the river, either directly or by later removal, were increased slightly, but relatively high variances preclude any detailed interpretation. The absence of direct removal of leaf litter and woody debris from the site by flood flows increases storage on the site. With this increase of material available on the site, decomposition also increases. The total amounts of litter produced remain the same, and a different equilibrium storage is reached by increasing the output through this alternative pathway in what is a negative feedback loop.

The key aspect of the spiral of carbon in the riparian area is the transport of material from the terrestrial to the aquatic ecosystem. While some redistribution on a floodplain occurs, no studies have emphasized this process or distinguished among tesserae that might act as a source or sink for organic matter. The degree of spiraling either within a terrestrial riparian zone or at the terrestrial–aquatic ecotone will depend on the nature of the tesserae. Mature forests at cutbanks will differ from pioneer communities on point bars; abandoned channels will differ from levees; topography, hydrology, and ecology will affect productivity, decomposition, and transport of both leaf litter and woody debris, and the specific spatial configuration of the tesserae will affect the summation of the process and the overall spiral or cascade. Decomposition and the transfer of carbon to the atmosphere will also need to be accommodated.

Landscape ecology

Principles and directions

Two of the general principles of landscape ecology address the cascades of matter and energy across the landscape (Forman and Godron 1986). One, the nutrient redistribution principle, states that redistribution among elements increases with disturbance. This principle, however, does not deal directly with the spatial configuration of landscape elements or with the topology of material flows. The energy flow principle states that the flows of heat and biomass across boundaries increase with heterogeneity. This principle, while spatial in nature, at first seems to be a tautology: given any conditions of energy flow, if more boundaries exist, more crossing of boundaries will result, and an increase in heterogeneity leads to an increase in boundaries. Boundaries themselves, however, do lead to increases in the rates of flow between elements. A restatement of these principles, meant to clarify certain aspects related to topology in the riparian situation, rather than to eliminate them, may be useful. A restated principle would hold that the rate of redistribution of matter or energy between landscape elements increases with the spatial proximity of the elements, the length of their shared borders, and the gradient (especially topographic for matter) between them.

Riparian landscape elements have notable edge and gradient characteristics. It is their particular location between upland and stream that creates this situation. The interaction between a riparian area and a stream is clearly one where flows across the boundary are important, and the landscape approach has been used, both implicitly and explicitly, by researchers in aquatic ecology who have tried to understand the terrestrial inputs to the aquatic system (Decamps 1984). Whole ecosystems depend on this input to the aquatic food-chain (Cummins *et al.* 1989). The relation between the general character of a watershed, e.g. the dominant vegetation, and some characteristics of streams has been studied for some time (cf. Likens and Bormann 1974).

Hynes (1975) described the relation of lotic ecosystems and the landscape of their drainage basin for large rivers. He noted that the importance of the landscape had not been recognized in many earlier studies This failing has since been corrected. Lotic ecology now uses a more spatial approach in research on the spatial heterogeneity of structure within the aquatic area (Meyer 1990). The most significant development in the study of the linkage between spatial patterns in lotic ecosystems and spatial patterns in their watershed has been the River Continuum Concept (Vannote *et al.* 1980). One example of the consideration of the longitudinal structure of the landscape is that of Tate and Gurtz (1986), who directed their attention toward prairie streams. They considered that prairie streams might function differently because their landscape structure differed, with the main input of leaf litter in higher rather than lowest-order reaches. They found differences in leaf decay between seasonal and perennial streams. Naiman *et al.* (1988*a*) have proposed a complement to the river continuum concept, a river mosaic concept in which reaches are considered as discrete patches with boundaries, and wherein the particular connection between the aquatic and riparian zones can be emphasized.

Schlosser and Karr (1981*b*) particularly noted the importance of the spatial distribution of riparian forest in the watershed for its role in affecting the cascade of sediment and nutrients. Although Omernik *et al.* (1981) found that the position of forest in an agricultural watershed did not have an effect on sediment yield, their study was limited and they identified several reasons why their results should not be generalized. Concepts in landscape ecology, as well as recognition of the shortcomings of regulations on point and non-point pollution sources in rivers, have led to efforts to characterize the cumulative effects of both these inputs and of the landscape structure of a drainage basin (Gosselink *et al.* 1990). Klock (1985) examined the cumulative effects of forestry practices on in-stream ecosystems. Childers and Gosselink (1990) noted the importance of the problem of cumulative impacts and the need for a landscape level approach. They proceeded, however, to a non-cumulative assessment, although they claimed that they were looking at the cumulative problem. A cumulative approach must look at the additive or multiplicative effects of impacts or mitigations as they accumulate over time or space, or both. Childers and Gosselink (1990) did not have any spatial data, they did not attempt to quantitatively link their temporal data to any impacts which were temporally accumulating in their basins, and they did not establish accumulation. Their study is an example of a claim of an approach, or 'buzzword', that has gained favor and funding from environmental agencies such as the sponsor of this project, the USA Environmental Protection Agency. In contrast, Johnston *et al.* (1990) have progressed in the direction of an assessment of the cumulative effects of wetlands on river water quality. They used a geographic information system to identify the position of wetlands in drainage basins and they found that this relative position had a major affect on water quality. The actual spatial configuration of the landscape may affect the ways in which the effects accumulate.

The intersection of geomorphology and ecology is one of the primary focuses of landscape ecology. Two aspects of this intersection have received some attention, i.e. the influence of the former on the latter, biogeomorphology (e.g. Viles 1988) and of the latter on the former, phytogeomorphology (e.g. Howard and Mitchell 1985), but most of the studies in these areas have not addressed spatial relations. Salo (1990) identified the geological and geomorphological processes that affect riparian ecosystems, and in his analysis it is implicit that it is the relation between geomorphology and ecology that gives new insights in landscape ecology. He particularly contrasts this approach with geomorphologically uninformed island biogeography. He presented a useful clarification of mega-, macro-, meso-, and microscale geomorphological processes that affect riparian ecology. He also showed the variety of spatial patterns of tesserae that can develop in a geomorphologically complex setting. In other settings, Castella and Amoros (1984, 1988) and Rostan *et al.* (1987) have showed that former riverine geomorphology may affect subsequent ecological development, and Butler and Malanson (1993) stressed that the particular spatial positioning of elements can affect lotic and lentic ecology. Elucidation of the effects of relative location of geomorphological and ecological landscape elements will be a notable contribution to landscape ecology. It would seem that an expansion of spatial analyses in biogeomorphology and phytogeomorphology would have been a natural development in physical geography, but apparently landscape ecology *per se* will be the impetus.

Spiraling as a concept needs to be rethought for river systems as a whole. It worked well for in-stream processes, but lateral and vertical transfers are more difficult to handle. Research on the Rhone (e.g. Dole 1983, 1984; Amoros *et al.* 1986) has emphasized a three-dimensional approach, but has addressed structure more than process. The dimensional problem for nutrient spiraling becomes especially clear when dealing with elements for which there is a large atmospheric transfer and storage. Nitrogen is especially important because the factor that allows riparian zones to successfully filter nitrogen in the long term is their capacity for denitrification. Carbon is likewise important because the amount of decomposition that can occur while material is stored on a floodplain will play a great role in how much carbon can be transferred into streams. In both of these cases the removal of the elements from the terrestrial–aquatic system needs to be included in any landscape level conceptualization of matter flows. Landscape ecology, by pointing out the importance of the spatial arrangement of tesserae, and thus complicating the in-stream scenario for which the nutrient spiraling concept was designed, leads us back to a more robust, but perhaps more unwieldy, approach of basic systems analysis using compartment models. In this case the specific spatial and locational features of compartments must be included. Junk *et al.* (1989) have emphasized the temporally and spatially heterogeneous nature of fluxes in the riparian zone and between this zone and the river. This pulse concept is a more sophisticated approach to the landscape than the river continuum and nutrient spiraling concepts because it is not based on assumed equilibrium conditions.

Conclusions

The features of the riparian environment as both a boundary between the terrestrial and aquatic zones and as a corridor need to be considered in examining the fluxes of energy and matter. The flow of water links all of the relevant processes, but it needs to be examined in three dimensions rather than as either in-stream flow or perpendicular over-bank flow. This complexity is clear in hydrology, but needs to be addressed in linking ecology and geomorphology. The importance of the mosaic nature of riparian areas will then be seen and concepts such as spiraling will be modified as a result.

6 · *Species dynamics*

A great blue heron flaps along before us, alights, waits, takes off again, leading us on. *(Edward Abbey)*

The dynamics of the vegetation of riparian areas should depend on the life and death of individuals and on regeneration. To the extent that regeneration represents an opportunity for new species to appear on a site, the dynamics can be considered as a cascade of genetic information across the landscape. This topic has received considerable attention in the general ecological literature under the topic of species dispersal (e.g. Sauer 1988), and I use the word cascade here only to draw a parallel with the relation between spatial pattern and process which has been emphasized in my discussions of cascades of matter and energy. One can consider a flow of species across the landscape, and in order to differentiate this flow from that of matter and energy, information is a general category which summarizes the distinctive aspect of life. Landscape ecology is fundamentally concerned with the movement of species (as well as energy and matter) across space at a variety of temporal and spatial scales (Forman and Godron 1986), but in this discussion I will primarily address the affect of riparian elements on the long-term, coarse-scale movement of species.

Species movement across space

The effectiveness of riparian zones as pathways for, or barriers to, the diffusion of genetic information is not known. Forman and Godron (1986) cited the role, but did not provide examples. Evidence is sparse, and much of the discussion must be hypothetical. The lack of evidence, however, may be due to the concentration of studies on other topics, rather than to any lack of effect. It would seem that several aspects of corridor structure, defined by Forman and Godron (1986), would have an affect on the effectiveness of riparian elements as corridors for the movement of species. First, the origin of the riparian elements as a resource patch is important because the resource will be tolerated or utilized differently by different species. Second, the spatial structure of the element is important. Curvilinearity or sinuosity may affect species movement. Breaks or nodes will either block or enhance species movement, and in sum connectivity will play a role. Narrows, or the constriction of a riparian element, will be utilized differently by different species. The overall pattern of the riparian zone, as it is controlled by the spatial structure of the general drainage network, may also determine the rate and direction of species movement.

Three aspects of the process of this cascade of information may operate. The first two are related to the linear configuration of the riparian zone, i.e. its role as a corridor or as a barrier. The width, continuity, and drainage pattern of the riparian habitat will play an important role in its function as a corridor, and the width of both the riparian habitat and the river itself will affect the dynamics as a barrier. The third general aspect is that of a more local shifting mosaic, or the local spatial dynamics of species and tesserae. Two other special topics are worthy of note: the role of riparian zones in exotic invasions and the landscape ecology of population genetics.

Evidence for past riparian species movement

Some of the evidence for landscape-level flows of species, in general, is to be found in research on the spatial dynamics of species at the end of the Pleistocene and through the Holocene. Gleason (1922) considered riparian corridors to be important features in the post-Pleistocene expansion of deciduous forest into the upper Midwest of the USA which had been occupied by ice, tundra, and spruce–fir forests. The overall movement of species in eastern North America has been documented through palynological analyses (e.g. Davis 1981). Constructions of isopols, or contours of pollen concentrations, have been successfully used to document the individualistic shift of species to the north and east that occurred in the Holocene. Actual data on species dynamics in the upper Midwest USA during the Holocene is not able to address the question of corridors or barriers, however. The sites in which pollen collects and is preserved are not able to spatially distinguish the pattern or location of species within a small region. Pollen collects in ponds and marshes, and may represent a spatially averaged vegetative composition. In contrast, plant macrofossils, i.e. leaves and seeds, do not disperse as widely, and represent a more local sample of vegetation. Riparian ponds, backwaters, or abandoned channels may collect local riparian fossils, and thus represent the species dynamics of these environments.

Data from riparian sites do not indicate how they may have operated differently because of their spatial configuration, but do indicate, however, that diffusion of species does take place through these environments. In eastern Iowa, the pollen sequence from Roberts Creek (Chumbley *et al.* 1990) and a preliminary pollen and macrofossil sequence from Mud Creek (Baker *et al.* 1990) indicate that mesic deciduous forest occupied these areas from about 9500 to 5500 BP. The pollen records contain information on the vegetation from both uplands and lowlands, and the plant macrofossils indicate the species composition of the floodplains and adjacent uplands during periods of strikingly different vegetation during the Holocene. At Roberts Creek, *Ulmus* and *Quercus* replace *Picea* c. 9000 BP, *Tilia* joins them c. 6000 BP, but they rapidly decline c. 5500 BP when prairie vegetation arrives; the oak returns after 3000 BP (Chumbley *et al.* 1990). This sequence at Roberts Creek clearly contradicts the history postulated by Webb *et al.* (1983). Farther east, at Devil's Lake, the mesic forest species declined at c. 5500 BP, but prairie did not enter; instead, a more xeric, oak-dominated forest arose (Maher 1982). The sequence at Mud Creek is quite similar to that at Roberts Creek (Baker *et al.* 1990). These data indicate that the riparian nature of these sites may have affected the flux of species across the landscape or at least the recording of the overall flux of species.

Bettis *et al.* (1990) reported the Holocene advance of *Carya illinoensis* into the upper Midwest along the Mississippi River. This species is found primarily in interior riparian forest sites in this region. They concluded that the edible nuts of this species were carried northward by human populations rather than by water, which flows south, or by animals, which would have moved too slowly. Unfortunately, they had only a single macrofossil nut of the species, and they were not able to distinguish the pollen of this species from those of other *Carya*. Although they considered almost all other species as strictly upland types, in fact several others (e.g. *C. cordiformis*) are among the dominants of the present riparian community, and because of their taste would not have been transported by humans.

Some definitions

It is important to be clear about terminology at this point. A variety of uses of the words dispersal and migration have been proposed. Without arguing about the history of these developments, I will use three words in specific ways:

Diffusion: the spread of a species into new areas. For example, diffusion occurred when plant species moved to more northerly ranges at the end of the Pleistocene (Davis 1981). Diffusion is the sum of dispersal events. It is not germane to worry about confusing physicists or chemists who may have a particular connotation for this term involving randomness, and in geography the word has a solid history of usage for the spread of ideas (cf. Abler *et al.* 1971).

Dispersal: the movement of a single individual to an area not previously occupied. For plants, or sessile animals, this occurs when propagules are disseminated. A seed moving to any site away from the parent is a dispersal event. Animal dispersal occurs when an individual moves out of its normal home range and either expands this range or establishes a new range. The normal home range may have been that used by the parent animals if the dispersal is by a juvenile.

Migration: the movement of animals between disjunct seasonal home ranges. Plants do not migrate (cf. Sauer 1988 for an alternative definition).

These distinctions are important because they have definite spatial components, and the principles of landscape ecology must be considered from a viewpoint in which these spatial components are distinct. Diffusion is clearly a landscape-level process. Dispersal may be a landscape-level process, but may also occur within a single tessera. Migration is also a landscape-level process.

Dispersal in riparian habitats

Dispersal and regeneration depend on the reproductive strategies (*sensu* Giesel 1976) and life history traits (or together the vital attributes; Noble and Slatyer 1980) of the species. The actual combinations of strategies found in riparian areas have not been documented, but some studies of

Figure 6.1 A hanging garden in the Narrows, Virgin River, Zion National Park, Utah.

aspects of the problem provide useful insights. The basis for much future work must rest, however, on established concepts of dispersal.

Dispersal spectra of riparian plant communities

Luftensteiner (1979), citing Braun-Blanquet (1951, p. 396), presented the concept of dispersal spectra. Alternatively, I have referred to the assemblage of dispersal types (Malanson and Kay 1980; Malanson 1982), building on Diamond's (1975) rules of assembly and Levin's (1976) three controls of spatial pattern in which they noted that dispersal would matter. The basic idea is that the coexistence of species in some plant communities may be due to their sharing of dispersal traits rather than to other shared traits, e.g. physiological tolerances or requirements.

I examined the distribution of dispersal types in hanging gardens in the Narrows canyon of southern Utah. This is an extreme form of riparian environment, wherein everything is an edge, but where the plants grow only in distinct isolated patches (Figure 6.1). The plant communities are quite small and cling to the sides of vertical cliffs where hydrological seeps have created a resource patch. I found that species were grouped by dispersal type: spore dispersing ferns (e.g. *Adiantum capillus-veneris*,

Cystopteris fragilis) and mosses dominated the smaller, more isolated, and most frequently flooded sites, while rarely flooded sites had a more diverse assemblage including the flowers *Aralia racemosa*, *Dodecatheon pulchellum* and *Mimulus cardinalis* and the trees *Acer negundo* and *Fraxinus velutina*. The most important factor was that the ferns and mosses were able to reach frequently flooded sites soonest, and so maintain their dominance as fugitives. Even this analysis could have been more spatially precise. I considered linear distances among sites along the bottom of a very deep and narrow canyon as the measure of relative location, but I did not consider the effects of sinuosity on wind flows nor alternative pathways that could possibly be taken by birds.

Weaver *et al.* (1925) speculated that the longitudinal and transverse structure of riparian vegetation in Nebraska was the result of a headward diffusion of species. They identified a longitudinal gradient of dispersal types, with anemochores at headwaters, and rodent- and bird-dispersed zoochores more abundant in lower reaches. In their model, initial invasion follows soon upon fluvial incision, and the transverse structure develops subsequently. This report is most illuminating in regard to the concepts that these noted ecologists brought to their study. The longitudinal and transverse structures that they described, however, do show definite associations among species and between species and landscape positions. Their observations of the spatial distribution of dispersal types deserve further investigation to determine if this is in fact the result of a spatial process or if their distribution is related to correlated morphological or physiological traits of the mature forms.

Actual studies of the seed rain on a site are difficult because of the various pathways of dispersal. Seed traps, which are effective for wind-carried diaspores, will not work equally well for water-borne seeds and animals may avoid them. Even for wind-dispersed species, the seed trap concept is problematic because the trap implies an endpoint of dispersal while that same point may have been a temporary stop for seeds in the absence of the trap.

Diaspore or propagule morphology

Types of diaspores have been distinguished by several botanists. Some have addressed the morphology of the diaspore directly, without inferring dispersal mechanism (Dansereau and Lems 1957). This method has the advantage of not prejudging the agent of dispersal, but is cumbersome because in this case it uses 10 categories. Van der Pijl (1982)

Table 6.1 *A categorization of dispersal types based on the agents identified by van der Pijl (1982) and the morphologies identified by Dansereau and Lems (1957)*

Hydrochores (dispersed by water)
Cyclochore (voluminous, loose spherical framework)
Sarcochore (juicy or fleshy outer layers)
Pterochore (appendages wing-like)
Sporochore (hard outer layer; small enough to be carried by a breeze)
Sclerochore (too heavy for a breeze, but light enough to be carried by wind)
Desmochore (appendages spiny or glandular)
Anemochores (dispersed by wind)
Cyclochores (voluminous, loose spherical framework)
Pterochore (appendages wing-like)
Pogonochore (appendages long, hair-like, or plumose)
Sporochore (hard outer layer; small enough to be carried by a breeze)
Sclerochore (too heavy for a breeze, but light enough to be carried by wind)
Zoochores (dispersed by animals)
Sarcochore (juicy or fleshy outer layers)
Desmochore (appendages spiny or glandular)
Others (e.g. anthropochores dispersed by people)

based his categorization on the primary agent of dispersal, e.g. wind, water, or animals. It is important to recognize that the diaspores of a single species may be dispersed by more than one agent. I try to combine the agents of van der Pijl (1982) and the morphologies of Dansereau and Lems (1957) in Table 6.1.

Hydrochores

It seems natural in an examination of dispersal processes in riparian areas to consider hydrochory (dispersal by water) first. Many of the studies of hydrochory have been directed toward long-distance oceanic dispersal (Sauer 1988). The dispersal of the coconut (*Cocos* spp.) has even entered into a form of popular culture, with its potential presence in medieval England being discussed in the opening scene of Monty Python's *Search for the Holy Grail*. Dispersal by rivers has received much less attention. Sauer (1988) pointed out, however, that establishment of riparian species can be closely related to the combined effects of river flow, including opened sites, moist sediment, and dispersal. Indicative of its potential importance in riparian areas, Skoglund (1990) found that hydrochory was the only source of seeds of species not already present on a riparian meadow site.

An interesting study of the process of hydrochory is the recent work of Nilsson *et al.* (1991*b*). They used small wooden cubes to simulate floating seeds. They dropped them in midstream, and then recorded if and where they were stranded along the river banks for 3.7 km. They found no association between current velocity, riverbank width, or substrate and the number of cubes stranded, but a negative association with the percentage cover of ground vegetation on the riverbank. They could not explain this effect; it could be due to the creation of a barrier by the vegetation which increased downstream flow, but this effect would depend on the particular structure of the vegetation. They compared the floating abilities of seeds of the plants found along the river with the numbers of cubes stranded, and found that sites that captured the most cubes tended to have species with seeds that floated longest.

Schneider and Sharitz (1988) reported the importance of hydrochory for the dispersal of *Taxodium distichum* and *Nyssa aquatica*, the two dominant species of riparian swamps in the southeastern USA. Mean floating times of 42 and 85 days, respectively, aided in the directional transport by water. Distances of nearly 2 km were covered by the diaspores from single trees. Deposition was nonrandom, with accumulation related to emergent vegetation and woody debris. The soil seed bank reflected this nonrandom distributional process. Although this study is most germane to the distribution by slow-flowing flood waters in this system, the authors noted that occasional higher flows potentially can scour the surface and transport diaspores for much longer distances.

Some problems exist in the study of hydrochory and the assessment of its importance. These problems are related to the importance of floating by the diaspores. Variations in floating time have been noted among species and this has been regarded as an important indicator of the potential success at dispersal by water. For riparian areas, however, only the evidence of Nilsson *et al.* (1991*b*) indicates that floating time may be important. Some inference has been made that once floating seeds sink, they are no longer viable (Nilsson *et al.* 1991*b*). In some cases this may be true, but I have examined floating among the acorns of various species of *Quercus* in Iowa, and found that some individuals will float with their cap, but not without. The loss of the cap does not affect its viability, but without this buoyant flotation device the acorn stops floating. It is also possible that diaspores which do not float can be effectively dispersed by flood flows. Seeds which do not float are not like stones: they are often just slightly too dense to float. Yet even stones can be moved downstream as bed load during flood flows. On a wide floodplain such a seed could be carried great distances during over-bank flooding. Only if the seed was washed into the channel or into a pond where it would stay submerged would its dispersal by water necessarily be fatal. Van der Pijl (1982) noted submerged transport, but cited it mainly for aquatic species.

Another problem for hydrochory is the prevalence of mixed modes of dispersal. Many diaspores that have been classified as having other means of dispersal can also float and/or be dispersed by flowing water. Sauer (1988) specifically noted that sites in the midwestern USA are commonly occupied in early successional stage by species that disperse via both wind and water. Thebaud and Debussche (1991) documented the rapid range extension of *Fraxinus ornus*, normally an anemochore, along a riparian zone in southern France, which they attributed to hydrochory. This species was able to advance at a rate of nearly 1 km yr^{-1}, but was restricted to riparian sites disturbed by flooding. Telenius and Torstensson (1991) concluded that the presence or absence of seed wings in the genus *Spergularia* may have evolved to use wind a a primary dispersal agent and water as a secondary one. Herrera (1991) reported that the diaspores of *Nerium oleander* float downstream and colonize streambanks, and that although their tuft of long hairs suggests an anemochore, wind dispersal will be limited because the pappus folds hygroscopically and the seeds are released during the rainy season. Along rivers in Iowa I have observed several species with specific adaptations for dispersal by wind being carried downstream after landing on the surface of the river. These include *Populus deltoides*, *Salix* spp., *Acer saccharinum*, and *Ulmus rubra*. I have observed great accumulations of the samaras of *A. saccharinum* along the strand line at point bars and have later observed seedlings in some of the same places which had fine-grained sediments and more moisture. Species commonly thought to be dispersed by animals can also be carried by water. I have already cited the acorns of *Quercus* species. While still in their husks the diaspores of *Carya* species float, but without husks they depend on transport by animals. Murray (1986) noted that species with adaptations to cling to animals could also be dispersed by running water (e.g. *Xanthium occidentale* in Australia). Van der Pijl (1982) also noted the potential for multiple dispersal pathways for species.

Anemochores

Many riparian tree species are wind-dispersed. The wind is a reliable agent and the development of special traits does not involve a major expenditure of energy. In fact, wind dispersal requires that seeds be relatively light, so energy investment must be minimal (cf. Ganeshaiah and Shaanker 1991). Wind-dispersed species are more often *r*-selected and are successful pioneers of newly opened sites with high ambient light and bare mineral soils (e.g. *Populus* and *Salix* species in many riparian environments worldwide); however, a notable gradient exists among species, and some interior canopy dominants such as *Ulmus* spp. are also wind-dispersed.

The general study of the process of dispersal by wind has notable problems. Some studies have examined the seedfall from an isolated, individual plant. The results of these studies indicate a 'footprint' of dispersal around the plant that shows a negative exponential decay with distance, although this might be reversed under the canopy itself (Johnson *et al.* 1981). The footprint is elongated in the direction to which the wind blows, and in theory is related to the wind-rose for the period of seed release. These studies are useful in describing the general form of the dispersal function, and in differentiating among species and dispersal types, but they do not give the true probability of dispersal to another site for a species living in a plant community. Multiple trees will greatly affect the wind speeds and turbulence, and in any case these studies are not able to document long-distance dispersal beyond the range of the collections which might occur during very high winds.

Another approach has been to calculate the rate of fall of a diaspore in still air and then to compute dispersal distance as

$$D = (\text{height of parent} \times \text{wind speed})/\text{rate of fall}.$$

Species with obvious adaptations for wind dispersal, such as pterochores or pogonochores, have lower rates of fall and so longer dispersal distances. This formulation would be accurate if winds were laminar, but they are not. Burrows (1986) detailed the aerodynamic properties of diaspores in relation to the physics of the problem, and turbulence in wind movement causes the measurement of the rate of fall in still air to be nearly meaningless. I have seen *Populus deltoides* diaspores floating in apparently still air with no observable rate of descent and some uplift. I have also observed seeds of other species, e.g. *Taraxacum officinalis* (dandelion), being lifted an order of magnitude higher than the tallest parent plant. Suffice it to say that estimates of dispersability based on such studies need to be inflated (cf. Peart 1985). Studies using seed traps in isolated locations would seem to be one way to get data on long-distance dispersal, but such studies can be time-consuming and, with negative results, inconclusive.

Zoochores
Mammals Van der Pijl (1982) distinguished dispersal among rodents, ungulates, bats, and primates. Squirrels and other seed hoarding species can be effective dispersal agents over short distances. Such means of dispersal require a significant investment of energy in order to attract the vector, and a significant loss to actual consumption. Some riparian tree species of *Quercus* and *Carya* in the midwestern USA can be dispersed by small mammals, but one species, *Carya cordiformis* (bitternut hickory), however, is not palatable, and one wonders at its dispersal. Wider-ranging mammals, such as deer, to which propagules can adhere with hooks or sticky surfaces (desmochores), can carry seeds even farther, but this means seems to be used primarily by herbaceous or shrubby species (Van der Pijl 1982).

Birds By flying long distances, birds can be very effective dispersal agents. Van der Pijl (1982) cited several categories of ornithochory, which include nonadapted species and accidental dispersal, but the specific adaptations and behaviors of birds are most interesting. Plant species have adapted fruits which are attractive to birds (as well as to mammals), but one of the most interesting observations is that by Darley-Hill and Johnson (1981; W.C. Johnson, personal communication) who noted that bluejays (*Cyanocitta* spp.) can carry several acorns of *Quercus* spp. for a few kilometers. Because of the behavior of the birds, edge sites are again most likely to have individuals arising from and leading to such successful long-distance dispersal.

Fish In Chapter 3 I cited Goulding's (1980) report that ichthyochory might be important in the Amazon basin. Earlier workers in this region documented fish eating habits and the contents of fish guts, and considered the possibility of ichthyochory in detail (Huber 1910; Kuhlman and Kuhn 1947; Gottsberger 1978). While fish eat seeds in many places (van der Pijl 1982), active use of the floodplain by fish during over-bank flooding may be necessary to make ichthyochory an important factor in riparian landscape ecology. This phenomenon is not, however, confined to the extreme flooding of the Amazon (Crance and Ischinger 1989). Distinguishing this pathway from others will be difficult.

Anthropochores
While we technically are mammals, people disperse seeds in different ways and over much longer distances. In some instances, the mechanisms may be similar to mammal dispersal, as in the case of movement with food. The potential dispersal of *Carya illinoensis* by prehistoric peoples in America cited by Bettis *et al.* (1990) would be one case. People are also responsible for notable riparian invasions, discussed below, arising from deliberate introduction of species from one continent to another.

Seed banks

Sauer (1988) reported on the importance of seed banks for riparian herbs in Missouri. He identified microdiffusion as a means to maintain these seed banks in riparian areas where spatially progressive erosion continually creates new sites. Schneider and Sharitz (1986) examined the role that a seed bank plays in flooded swamp and a bottomland floodplain forests of *Taxodium distichum* and *Nyssa aquatica*. The nature of the seed banks fluctuates through the year with seed inputs and with flooding events. While Finlayson *et al.* (1990) found the seed bank to be important for regeneration on floodplain sites in northern Australia, seed banks may be a more important vital attribute for still wetlands such as prairie potholes (e.g. van der Valk 1981) than for riparian sites subject to flowing water.

Corridors

The function of a riparian corridor as a conduit for species movements may occur through active transport by the fluvial process in what would normally be considered as passive dispersal, such as when a seed is carried downstream and deposited on a distant floodplain. Alternatively, the riparian system may play a role as a passive corridor for the more active dispersal of animals, or through them of zoochores, or may channel wind and anemochores. Forman and Godron (1986) identified width as the most important feature of corridors for species composition. Width may also be the most important feature for species dynamics, but connectivity, sinuosity, and network pattern will also play a role.

Edges

The edges of the riparian zone provide unique dispersal pathways because of the spatial arrangement of accessible sites. Accessibility is a key factor in species dynamics (Kellman 1970), and edge sites are particularly accessible to certain types of organisms dispersing across the landscape. Riparian edges, in contrast to the more commonly studied edge features such as those at agricultural fields or clear-cuts, appear to be continuous and extensive. Riparian edges extend from coastal areas to the heart of continents (e.g. the Nile's >6600 km length). A notable feature of the edges of many riparian sites, however, is that in contrast to the human perspective, from the viewpoint of the species, the edges are not continuous. That is, in meandering rivers the edge on any one side changes from a point bar site to a cutbank site, with a zone of transition between. The riparian element is continuous, but the tesserae vary. The sinuosity of the riparian element plays a role in the arrangement of the tesserae and the distances between them. The distance from one point bar to the next, even crossing the river, is not necessarily small. This spatial separation should be considered when identifying dispersal pathways.

Edges are particularly accessible to anemochores, or wind-dispersed species. Point bars are often colonized by *Populus* and *Salix* species, and as noted in Chapter 4, the dispersal event may play an important role in the internal structure of the riparian plant community as well as in the eventual biogeomorphological development of the riparian landscape element. Riparian edges are also accessible to some birds and bird-dispersed seeds. Darley-Hill and Johnson (1981; W.C. Johnson, personal communication) have documented the use of edges by bluejays and their propensity to bury *Quercus* acorns, which they carry long distances, at the edges of woodlands. In Iowa I have observed more regeneration by *Quercus* along cutbank edges than anywhere else in the riparian zone. The riparian edge sites are also accessible to hydrochores. Point bars are often the site of deposition of many seeds. The samaras of *Acer saccharinum* and the parachutes of *Salix interior* can float, and strand lines of these seeds can create dense linear stands of seedlings. Dietz (1952) described this phase of dispersal and establishment for *Salix* spp. in Missouri, and we found this phenomenon to be widespread in Iowa (Craig 1992) (Figure 4.5).

Interiors

Interiors of riparian zones are probably less active pathways for plant dispersal. As in the interior of any vegetative stand, wind is reduced and so anemochore dispersal is likely to be limited. Stands of *Acer saccharinum* and *Ulmus rubra* in Iowa produce dense deposits of their seeds directly beneath the parent plants. Dispersal by birds and mammals may also be limited. Although birds and mammals do use and move seeds in these interior sites, interior animals tend to have smaller home ranges than those of edges, and so dispersal distances are likely to be less. Hydrochory may be an important factor in the dispersal of interior plants. Over-bank flooding can carry considerable amounts of large woody debris and litter within a riparian interior, as discussed in Chapter 5, and seeds are also moved in this way. Most of the studies of litterfall cited in Chapter 5 distinguished reproductive material, although this would have included inflorescences. Sloughs or abandoned channels, the primary pathways for water in this area, will thus also be the primary locations for hydrochore dispersal.

Fragmentation

Hanson *et al.* (1990) adapted a forest stand simulation model to study the effects of dispersal and forest fragmentation in a riparian corridor. They found decreases in diversity and changes in species composition in more fragmented landscapes. Species with less effective dispersal pathways, especially those dispersed only by gravity or mammals, were most impacted by this spatial effect. Open fields that can isolate fragments were effective barriers to dispersal by small mammals, such as squirrels, which would carry *Quercus* and *Carya* species. This model, however, greatly simplified the spatial structure of the riparian element, but it could be modified in order to examine the effects of sinuosity or the existence of nodes.

One aspect of fragmentation is the increase in edge. Along riparian areas, however, the new edge is not the same as the river edge, which is topologically in the same position, and in fact fragmentation leads to the reduction in this type of natural habitat. In agricultural areas, the practice of farming directly up to the river's edge where possible illustrates this complete elimination of the edge habitat (Figure 6.2). While not directly breaking the corridor, human activities also partially fragment riparian lands by changing all but a very thin corridor along the river (Figure 6.3). In this case, only edge exists. If only edge sites are accessible, only edge species can disperse to these sites or at a coarser scale only edge species can use these sites as pathways for diffusion across the landscape, but for interior species that are diffusing across the landscape, these edges are barriers. For example, a seed of shade-intolerant *Cornus florida* that lands in such a location may produce a tree which will produce seeds

Figure 6.2 Fertile farmlands are occasionally cultivated as close to the edge of the river as possible; White Breast River, Iowa.

which will be carried further along the corridor in the next generation, but a seed of shade tolerant *Carya laciniosa* may never germinate. The actual operation of these edges as conduits or barriers will depend directly on the means of dispersal. For species dispersed by mammals, an edge may be useable as a pathway for the animal, while it would not be suitable for establishment, and in this way interior species can be transported along this edge. General rules may not be possible to apply to any specific cases, and instead each species and its relations with others may need to be considered in evaluating the effect that landscape structure will have on dispersal and diffusion processes.

Network structure

Network structures have been defined and analyzed for the movement or diffusion of many things, e.g. ideas, across landscapes or regions (Abler *et al.* 1971). River systems have several distinct network patterns on the landscape (cf. Forman and Godron 1986, p. 149). While quantitatively and topologically different, these patterns are all examples of branching structures. Common elements of branching structures have

Figure 6.3 In some cases where farm fields do not reach the river's edge, a very narrow strip of riparian vegetation is left.

been examined (e.g. Woldenberg and Horsfield 1983), but from the perspective of riparian species movement, key elements are the directions of river flow and the effect of the network structure on wind and animal movement.

When the environment changes do riparian species diffuse (or migrate) under particular constraints? If stream courses run perpendicular to the direction of diffusion then species must disperse between drainage basins and will face the problems associated with dispersal to, and establishment on, any isolate. If the direction of dispersal is parallel to the stream then the riparian zone may act as a corridor. But if species are more often carried downstream and the direction of diffusion is upstream, then diffusion will be especially difficult. For species restricted to a narrow riparian corridor, the constraints on diffusion will be greater than for a widespread forest. Winds perpendicular to the corridor will tend to carry light or winged seeds out of the riparian zone. Heavy seeds such as acorns (*Quercus* spp.) or hickory nuts (*Carya* spp.) can move only downstream in flood flows and their dispersal in the upstream direction would be tied to the movement of mammals and birds.

Hydrochores in a riparian corridor are dependent on the downstream direction of flow. The riparian network is a one-way system, and all hydrochores will move downstream. This truism becomes of interest when examining the effects of hydrochory at a landscape or regional scale when the riparian network includes a significant longitudinal environmental gradient. Such gradients were discussed in Chapter 3; they include latitudinal, elevation, and moisture gradients. The degree to which hydrochory can affect the pattern of species distributions along the longitudinal structure of the riparian zone has not been studied. Notable gradients occur where river networks cross climatological regions: e.g. the Mississippi River system, crossing a gradient of temperature from cold to warm; the Colorado River crossing a gradient from wet to dry; the Brazos River in Texas crossing from dry to wet; or the Mackenzie River crossing from cold to very cold. Such rivers along latitudinal or moisture gradients occur worldwide. All rivers flow downhill, but elevational gradients can be significant where a river crosses biomes most directly related to elevation and analogous to latitude. In this case the analogy is directly to equatorward-flowing rivers. A case where a poleward-flowing river and a significant elevation gradient are combined would be an interesting contrast.

For anemochores, the orientation of the riparian element relative to the wind is important. If the riparian zone is perpendicular to the dominant wind direction, then the seedfall footprint for a plant might fall outside of the riparian zone in inhospitable territory. Alignment of the riparian corridor with the direction of wind will maximize the effectiveness of anemochory. Winds do not blow from one direction constantly, of course. The variation of the wind direction and speed, embodied in a wind-rose relative to the network, will determine its effect. The effect on specific species must be narrowed down to the time of seed release.

A notable feature of the branching structure of riparian networks is that they tend to funnel together in the downstream direction, and, conversely, all diverge in the upstream direction, which may be important for zoochores and the dispersal of animals themselves. For species moving upstream, divergence may lead to smaller populations and spatial isolation. Aspects of a funnel effect (*sensu* Forman and Godron 1986) may operate. For downstream movement, separate groups of a metapopulation, or formerly separate species, may merge. Competition may increase. The way that such a funneling or diversion may affect the behavior of migrating or dispersing animals is also a matter of speculation. These potential effects of pattern on process need investigation.

Barrier effects

Conversely, riparian environments and their streams may be barriers to dispersal and diffusion across them. The primary barrier effect presented by the riparian area is the transverse width for species that would be dispersing across, rather than along, the river's course. Both the riparian zone and the river itself must be considered. This transverse barrier changes, however, in the longitudinal direction. The stream, and often the riparian zone as a whole, becomes wider as one moves downstream. The seeds of plants and many animals can easily cross a first-order creek, but the probability declines for wide rivers and riparian zones, and wide rivers can be nearly absolute barriers for terrestrial mammals.

The riparian zone

The width of the riparian environment on both sides of the stream, and not including the stream, is probably a barrier to the diffusion of plant species not adapted to these conditions, but may not be a barrier for many animals. The barrier effect is probably most pronounced where wide floodplains with high levels of moisture preclude the effective establishment and success of upland species. A possible example of this effect can be seen in the range maps of trees in the USA. Little (1971, 1977) mapped these distributions, and the Mississippi embayment stands out distinctly in them, as discussed in Chapter 3. One feature on these maps is a number of species whose range is confined to the east of the embayment and whose western boundary coincides closely with it. It may be that these species do not exist to the west of the embayment only because they have not been able to diffuse across it, nor disperse across it in a single step.

Anemochores will be affected if they cannot blow over the riparian zone. Upland species that disperse via wind are probably not suited to establishment, reproduction, and renewed dispersal in the riparian zone. For zoochores, long-distance dispersal by birds is likely to be effective, but other modes are not, in part because of the river itself.

The river

The river itself, if wide, may be a barrier for plants or animals. Some types of seeds which do not have special adaptations for dispersal by wind or animals, and which do not float or which are damaged by inundation, may not be able to cross rivers. Given the number of plant species that do not meet these criteria, animals may be more limited by the rivers themselves, and here velocity may also be important. The effect of the barrier will be seen differently for different animal species. For birds, only those unwilling to fly across an extensive open space might be limited. For mammals, the limitation may depend on their ability to swim. In discussing species dynamics in landscape ecology, Forman and Godron (1986) reported a study by Storm *et al.* (1976). They reported that of six foxes (*Canis vulpes*) tracked for six days, one took it upon himself to swim the Mississippi River near Anoka, Minnesota, probably using islands but with minimum distances of up to 82 m, for no apparent reason. Two other foxes swam across another, 55 m wide, river. These foxes, however, did not travel along the riparian zone or any other corridor. This same river, slightly to the south, is wider and differences in morphological traits of fox populations on either side have been found. In this area it is much less likely that a mammal the size of a fox would regularly swim across this channel, and so that it might be a barrier to any of the three aspects of species movement. While the same channel is often frozen in the winter, foxes remain close to their dens at this time. It would be interesting, however, to determine the effects of a given barrier on gene flow, given different ethological and population characteristics, such as might differentiate predators which might cross a kilometer of open snow and animals such as foxes or squirrels which probably would not. At a larger spatial and temporal scale, Colyn *et al.* (1991) hypothesized that although the Zaire basin may have served as a Pleistocene refuge for primate taxa, the river would have acted as a barrier. Their maps of primate and plant endemism support this river barrier concept.

Invasions

In a relatively recent book on invasions of North America and Hawaii containing 16 chapters by different authors, almost no mention of the spatial structure of the landscape is made in explaining invasions (Mooney and Drake 1986). Species and site characteristics are given a thorough review, but even the basic tenets of island biogeography seem to be lacking. Landscape ecology may provide a useful basis for evaluating invasions in general, and in riparian areas the simplified spatial structure may be an advantage. Riparian areas in the southwestern USA have been well studied with respect to a few exotic species. Complexity will arise, however; E. Tabacchi and A.-M. Planty-Tabacchi (personal communication) report that 400 (!) exotic species have been recorded in an ongoing study of the Adour and Garonne Rivers in France.

In the southwestern USA three genera of trees have been successful in invading riparian environments (*Eleagnus*, *Eucalyptus*, *Tamarix*). The most successful is the phreatophyte tamarisk. Graf (1985) noted that although earlier workers identified several species (e.g. *Tamarix aphylla*, *T. galicia*, and *T. pentandra*) they are now lumped as *T. chinensis* (cf. Baum 1967). *Eucalyptus* species and *Eleagnus angustifolia* (Russian olive) have also been successful, but they are not confined to riparian zones and perhaps not as aggressive in their invasion of them. Several studies described the processes by which *Tamarix* colonizes areas; some have documented rates of spread in specific river systems (Clover and Jotter 1944; Bowser 1960; Christensen 1962; Horton 1964, 1977; Robinson 1965; Harris 1966; Baum 1967; Warren and Turner 1975; Graf 1978, 1982).

Christensen (1962, 1963) reported on the original establishment and spread of *Tamarix* and *Eleagnus*, respectively. Primarily using historical information and older botanical records, he documented peak expansion of *Tamarix* along the Colorado and Green Rivers (as well as the shores of two large lakes) in the 1935–55 period and noted ongoing expansion along streams of the Wasatch Mountains (Christensen 1962). *Eleagnus* was observed as a strongly, but not exclusively, riparian invader (Christensen 1963).

Harris (1966) reviewed the status of *Tamarix* and attributed its spread to the construction of reservoirs, although he noted that natural flow regimes in the Southwest were ideal for its invasion. He provided a map of *Tamarix* coverage illustrating the nearly continuous distributions along most of the major rivers in the Southwest. It would be interesting to investigate where the distribution is broken along sections of rivers such as the Salt and Verde and, at least in 1961, through the Grand Canyon.

Because these invading species are phreatophytes they have altered the hydrology of the area, and because of their growth form they have altered the geomorphology of the area; as a consequence, they have been studied intensively (Horton 1972). Turner (1974) described changes in the vegetation along the Gila River. Maps from 1914, 1937, 1944 and 1964 show the major expansion of *Tamarix*.

Graf (1988) mapped *Tamarix* coverage in the Salt River, Arizona, channel area from 1900 to 1979. The species advanced until 1941, when it began a retreat attributable to increased drawdown of groundwater. These invasions have led to severe management problems, especially in respect of the alteration of the water balance through transpiration. Whether or not the spatial structure of the riparian zone in the Southwest is a factor in the species movement is not clear. Graf (1988) reported on the importance of new sediment, but patterns have not been analyzed. Howe and Knopf (1991) also reported that a changing hydrological regime led to success of both *Tamarix chinensis* and *Eleagnus angustifolia*, and noted that the change was leading to the demise of *Populus fremontii* in their study area in New Mexico. W.L. Graf (personal communication) has also hypothesized that climatic change could be an important factor in the original invasion and subsequent dynamics of *Tamarix*. The same phenomenon has been observed more recently in arid Australia, where Griffin *et al.* (1989) found a link between source areas and specific target sites for the invasion of *Tamarix aphylla*.

Knopf (1986, 1989; Knopf and Olson 1984; Olson and Knopf 1986*a*,*b*) has examined the invasion of riparian areas by *Eleagnus angustifolia*. He reported that its introduction was in colonial times and that it has spread in part because government agencies continue to promote its planting by selling seedlings at less than market value. In riparian areas *E. angustifolia* has been used by certain animal species, but it does displace native riparian trees and thus has a negative impact on other wildlife. Its growth may in fact widen riparian areas, leading to an alteration of their landscape structure.

Beerling (1991) reported on the invasion of riparian areas in Britain by *Reynoutria japonica* (Japanese knotweed). He identified the flood dispersal of rhizomes as the key agent of spread, because the species cannot reproduce by seed in Britain. Moreover, the type of land use affected its success, particularly the positive effects of disturbed sites such as spoil and construction areas. No spatial relations of these sites were reported, but it is likely, given the short length of the rivers considered and the aggressiveness of this invader, that none would be found.

In Australia, several European and American species have invaded along rivers. *Nicotiana glauca* and *Ricinus communis* have arrived from Europe, and *Argemone ochroleuca* and *Mimosa pigra* came from the Americas. The latter spread rapidly along rivers in northern Australia following its introduction in 1947. Dansereau (1964) commented that New Zealand riparian areas have been so greatly invaded by exotics that the native vegetation could not be determined. Similarly, du Preez and

Venter (1990) recorded many exotic species in riparian communities in South Africa.

These invasions are landscape phenomena. Dispersal is the key spatial process; the riparian zone provides the unique spatial pattern and people account for initial accessibility. Concepts of landscape ecology run through many of the studies of these invasions. A true landscape study linking structure and process, however, is yet to be done.

Landscape ecology

If the continual reproduction (in the dialectical sense) of the landscape is at the heart of landscape ecology, then species movement is the pulse of the landscape. From the relatively quick shifting of populations or individuals as a river alters its course to the slow diffusion of species through a riparian landscape, it is primarily the distribution of individuals and of species, i.e. of genetic information, that defines landscape structure, and it is the change in this structure – in these distributions – that defines the essential landscape process. The process of landscape reproduction by species movement has a component at every level of the ecological hierarchy, three of which are particularly important: genetic information at the population level, relative abundance at the community level, and spatial pattern at the landscape level.

Mosaic shifts within the riparian element

Plant communities have been characterized as a shifting mosaic (Whittaker and Levin 1977). This process of shifting occurs at a number of spatial and temporal scales, and the scale that one examines will in part determine the relation between the pattern and process that is observed. In riparian areas this difference can be seen in the relative importance of the landscape and internal structures in affecting the temporal dynamics of species. The internal structure will have a greater effect on more spatially and temporally limited time scales as in gap dynamics and succession. In Chapter 4 I discussed the progressive nature of geomorphological processes, such as those associated with meander migration, on the internal structure of riparian vegetation. In these cases the edge sites move across the landscape and those species occupying edges, on both point bars and cutbanks, must move with them. Neither the dispersal onto an expanding point bar, nor the dispersal back into interior forest as a cutbank erodes the edge, have been studied. Dispersal onto point bars or mid-channel bars from unspecified sources have been documented (Walker *et al.* 1986). Species movements need to be understood in order to fully conceptualize the riparian mosaic proposed by Kalliola and Puhakka (1988) and Naiman *et al.* (1988*a*).

Competition for space by dispersal

Let us for a moment consider a broader, evolutionary version of competition than defined by Grime (1979) or Keddy (1989) wherein a species is able to exist because it does something, anything, better than other species in the region. What we may see is not active competition, but the 'ghost of competition past' (Connell 1980), and in fact we may be focusing on the losers of past competition in the narrower sense. Most studies of competition between individuals or between species focus on the appropriation of resources, e.g. light, water, and nutrients. Some of the most interesting work addresses interference competition, where individuals interact directly; the clearest examples are among terrestrial animals, but examples are known for plants. Among plants, however, it has been easier to document, perhaps because it is most prevalent, exploitation competition wherein the individual to first sequester the resource wins. In some cases space occupied has been used as a surrogate for resources (cf. McConnaughay and Bazzaz 1991), and this mode can be called scramble competition for space. It is the way in which temporal gradients are divided by fugitive species, which temporarily at least win the scramble. One aspect of scramble competition that I do not wish to consider here is the growth of plants already established. This scramble is most notable in communities where many individuals resprout following fire and grow to occupy the initially empty above-ground space (Malanson 1985*a*). Conversely, following a disturbance or the creation of a terrestrial site (e.g. point and channel bar development) an important part of the scramble is initial dispersal and establishment. Species may have unique niche space based on their ability to disperse quickly and establish in conditions when direct interference or exploitation competition are minimal: this is the regeneration niche (Grubb 1977).

Petraitis *et al.* (1989) illustrated that differences in the immigration and extinction of species in patches could explain the maintenance of higher species diversity at intermediate frequencies of disturbance. This insight is directly connected to landscape ecology. Because of the spatially progressive nature of riparian mosaics, where edges and tesserae move around the landscape both continually and in discrete steps, they are an excellent place to examine the role of dispersal in defining the regeneration niche of species and thus in maintaining regional, gamma, diversity.

Accessibility

One key aspect of competition for space is accessibility (Heimans 1954; Kellman 1970). The existing landscape determines from where species can disperse, and accessibility is directly related to the corridor and barrier effects of riparian landscapes. The idea of accessibility emphasizes that a source as well as a target must be considered. Work in island biogeography that emphasized stepping-stones made this point (MacArthur and Wilson 1967). Studies of dispersal or diffusion often fail to consider sources explicitly; some studies, however, implicitly recognize accessibility. Cargill (1988) found that competitive success in colonizing disturbed riparian sites was related to accessibility, and Rushton (1988) found that a case of succession was determined by seed sources. M. Izard (personal communication) found that after nearly complete destruction of riparian vegetation by a flood along the Tech, surviving fragments of *Populus* served as sources for recolonization. Accessibility also affects animals, as shown by Szaro and Belfit (1986) for the lack of herpetofaunal colonization of a riparian island created behind a dam.

For one riparian area some interesting speculation has been made. As noted above, several authors have documented that rivers flowing across the Great Plains of the USA had very little woody riparian vegetation in the nineteenth century (e.g. Crouch 1979*a,b*), and Ranney and Johnson (1977) found source areas to be important among riparian and other forest islands in South Dakota. Currier (1982) noted that pioneer farmers on the Great Plains were encouraged to plant trees as shelter belts, and often planted *Populus* and *Salix* species, the two genera that came to dominate in the great expansion of riparian woodlands in the twentieth century. In addition to changing hydrological conditions that favored the establishment of these trees, their introduction to sites that could then serve as origins for dispersal to the riparian zone may have been important.

Related topics

Interspecific dependence In addition to competition, other interspecific interaction is important. When species diffuse, their new environment must contain all of the components necessary for their continued survival. Most specifically, a move by a species must be accompanied by any and all of the species on which it depends, i.e. hosts or obligate mutualists. For plants, one obvious mutualist may be pollinators for those species pollinated by animals. While many species may be pollinated by more than a single species, at least some pollinator must be present in the new environment in order for diffusion to be successful. I have found only one study that deals with the pollination of a riparian plant species. Herrera (1991) found that successful seed production by *Nerium oleander* is limited by pollen and that insect pollination, on which the species depends, is rare but, because the species blooms when little else in its Mediterranean environment is active, sufficient.

Genetic landscapes Rather slight difference in accessibility or gene flow can lead to differing gene frequencies along such a continuous environmental gradient (Endler 1977). Thus heterogeneous landscapes can affect the distribution of alleles within a species as well as the distribution of the individuals of the population. This genetic geography is becoming increasingly important as natural landscapes are fragmented by human activities and the conservation of genes and genetic diversity is seen to be an important part of biological conservation. Gene flow in populations has always been regarded as a significant component of genetic diversity. Endler (1977), in his book on the genetic geography, identified 80 of his references cited as concerning dispersal. Key work in this area would seem to include that of Dobzhansky and Wright (1943), Birdsell (1950), and Johnson and Heed (1975). The effects of spatial heterogeneity are still an important area of investigation in population genetics (Hey 1991; Vickers 1991). The combination of the spatial structure of riparian zones and the development of molecular techniques for genetic analysis provide opportunities for empirical tests of the ideas about space in population genetics which have been largely theoretical or developed for *Drosophila*. Some interesting studies on the population genetics of fish should provide a model for this investigation (e.g. Hamilton *et al.* 1989).

Conclusions

Species movement is an area where a landscape perspective has the most to offer to ecology and yet it is an area for which little basis for development exists. Study of such movement is difficult either because of the intensive observation needed to examine dispersal events or the spatially and temporally extensive observations needed to examine diffusion. Landscape ecology has at least provided a set of ideas that should allow better conceptualization of the problem and lead to better hypotheses. As discussed in Chapter 7, computer simulation models, in addition to field studies, can provide new insights at the landscape level.

7 · *Organizing the landscape*

It majestically performs its double function of flood of war and flood of peace, having, without interruption, upon the ranges of hills which embank the most notable portion of its course, oak-trees on the one side and vine-trees on the other – signifying strength and joy. *(Victor Hugo)*

Landscape ecology provides a perspective on both environmental structure and function. The structure of a landscape is reproduced continually through environmental functions which are controlled by the existing structure. Topology matters, but the vast number of interactions in, and the history of, landscapes makes each one unique, and thus place, as well as space, is important (Malanson 1987). Because of the unique character of landscapes the derivation of specific and generalizable scientific principles is difficult, but even the less specific principles that are now being proposed and tested by landscape ecologists are noteworthy. The importance of spatial scale for the organization of ecological information is one such area (e.g. Baker 1989; King 1991; Turner *et al.* 1991), and the recognition that at landscape-scales new ecological principles may emerge is stimulating. In concluding, I review the concepts presented earlier from the perspective of landscape reproduction, and I address some problems of studying landscape ecology. I specifically consider simulation modeling and geographic information systems as tools, and I use the context of understanding landscape-level biodiversity as a focus for this discussion.

Reproduction of the landscape

Landscape structure

Riparian environments have two major patterns or structures which can be considered as gradients. The longitudinal gradient, along the length from first-order streams to high-order rivers, is characteristic of their regional location. The transverse gradient, perpendicular to the channel, depends on local conditions. These structures develop from and in part determine the functional aspects of these environments in their processing of energy, matter, and species.

The longitudinal gradient in the riparian habitat depends on coincident gradients in the physical environment with which the biota interact (e.g. Zimmerman 1969). This gradient is distinguished by climate and represents a continuum of species distributions between cold and warm, or dry and wet, or in mountainous areas includes both types of climatic gradient. Hydrological conditions also change along this gradient as a river changes from intermittent to perennial or from influent to effluent. Geomorphological changes also coincide with the gradient, and these vary from erosional to depositional zones, with a dynamic equilibrium or transportational zone between. The importance of this longitudinal gradient is often ignored because of its relative stability. Another reason for the paucity of studies is that they would require many study sites spread across a continent, and so the logistical problems would increase greatly. As yet unstudied differences in these gradients among different rivers within and among regions might yield useful information.

The transverse gradient exists among communities and primarily is a response to the hydrosystem; it has received more attention. These community gradients and differences have been described and correlated with environmental variables in many areas. Different hydrological regimes create different conditions that are advantageous for establishment and growth of species which differ in their adaptations to these conditions. Thus, on a floodplain, there are gradients of establishment on new substrate, of physiological tolerance to flood duration and frequency and water table depths, and of competitive advantage. Because the paradigm of species adaptations to an environmental gradient deliberately ignores the actual distribution of species on the landscape, the spatial structure of these distributions, and any import it may have, it has largely been left for future study.

Landscape functions

The flows of energy and matter through the riparian habitat, including the temporary cycles within it, are controlled to a great degree by the hydrological regime. The importance of the landscape is obvious here because it is the source of the water. Drainage basin hydrology and landscape ecology address approximately the same spatial scale, and in both topology is critical. The relations among hydrology, fluvial geomorphology and riparian ecology have been studied for decades, but only in limited spatial settings, and landscape ecology is now providing a perspective that can integrate detailed knowledge of ecological processes with the already more spatially explicit science of hydrology. The recognition of the importance of the overall structure of the drainage basin and of the riparian zone within it should yield some useful directions for future study. In particular, the longitudinal gradient in discharge (increasing and becoming less variable in influent streams while decreasing and becoming more variable in effluent streams) should, in parallel with the River Continuum Concept (Vannote *et al.* 1980), generate new hypotheses.

While specific patterns of organisms and the flows of energy and nutrients have been given more attention in landscape ecology, the flow of species seems to be the heart of the paradigm. Both dispersal and diffusion can be studied in relation to the existing landscape structure. Species must start at some position that defines the ecological structure at that point in time. Sites which are accessible and safe must exist in a clear topological relation to the existing organisms. The processes of dispersal, growth, and renewed dispersal are fully conditioned by the topology of all the landscape functions and structures that I have discussed, and the flow of species across the landscape is distinct in the riparian zone. The topology is clearly that of a network, and so the ideas of connectivity, nodes, conduits and barriers are relatively simple but not necessarily easy to define and operationalize for real places. The spatial dynamics of species depend on the temporal changes in the spatial distribution of the physical structures and functions, and also on the simultaneous structure and dynamics of other species and individuals.

Reproduction of a system

The idea of reproduction of landscapes may seem alien. Reproduction is a loaded word in Marxian social science, in that social structures are continually reproduced by social forces in a specifically dialectical struggle. In landscape ecology, the word can usefully be borrowed. Landscape ecology similarly provides a conceptual, not an operational, model for the synthesis of spatial structure, morphological structure, and function in riparian ecosystems. The interacting dynamics of hydrology, geomorphology, and ecology reproduce the landscape. In that the processes are continual one can say that they are dialectical, but the number of processes may require a broader perspective or definition. The landscape structure will exert some control on its own reproduction, but the interaction in microclimate, nutrient cycling, erosion and deposition, and flood flows, coupled with continuing human pressures, presents complexities for which adequate models have yet to be developed. Furthermore, although processes are continual, time-lags in feedbacks, and thus nonequilibrium conditions, are evident in many environments (e.g. Malanson and Butler 1990; Kupfer and Malanson 1992*a*), and disequilibrium may affect most ecological and geomorphological structures and functions (DeAngelis and Waterhouse 1987; Malanson *et al.* 1992). Disequilibrium, a factor related to the historical uniqueness of places, can in fact lead to deterministic chaos (cf. Malanson *et al.* 1990, 1992). For example, May (1976) noted that increasing the lag time in simple population models leads to a near infinite number of solutions.

Forman and Godron (1986) presented the most general concept of landscape ecology: that spatial structure should matter. They highlighted the basic concept with some suggestive and tantalizing hints about how spatial structure might operate. The quantification of descriptors such as porosity, connectivity, and patch size, which can be related to structural or process variables, are an important first step in hypothesizing relations among variables and then testing the hypotheses. In very complex situations, however, the type of hypothesis testing common in much of science will prove intractable. For example, in a statistical test, when the variance attributable to any one variable is very small, it will take many cases before its significance can be really determined, and the possibility of multicollinearity and of other failures to meet the requirements for accurate statistical testing will mean that the order of importance of variables and the true power of the tests performed can not be assessed. Developments in statistical analyses are improving the ability of researchers to use these techniques, and more specific techniques have been reviewed by Turner *et al.* (1991). Their review provides a good basis for further investigation and assessment of future research directions.

At the present level of landscape ecology, statistical analyses alone, including spatial statistics, will best be used to outline the broadest areas of concern and serve as an inspiration for new and more insightful hypotheses. I would cite some of my own work as an example. In one study, we were able to infer that the differences in species composition among riparian sites was affected in part by the dispersal morphologies of the species examined (Malanson and Kay 1980; Malanson 1982). We were not able, however, to determine how these species might respond to changes in the system in any predictive way, nor were we able to document or clearly understand the actual dispersal pathways used by all of the individual species. In another study (Rex and Malanson 1990), we found that the shape of remnant riparian forest patches was related to human activity, river sinuosity, and valley width, but we were not able to distinguish how these forces interact in determining the shape of single patches, nor were we able to explain the lack of influence that additional within-patch edges, such as ponds, had on the shape index.

Work summarized in Turner and Gardner (1991) provides a basis for future quantitative approaches in landscape ecology. A more useful direction would be to incorporate the basic principles of landscape ecology into a framework of systems analysis. This approach will enable the investigation of the mechanisms by which landscapes are reproduced. From a perspective in environmental science, each iteration of a systems model reproduces the system.

Systems models and modeling

Modeling strategies

Three types of systems models will need to be combined in order to achieve this synthesis: hydrodynamic, geomorphological, and ecological models. Models in these areas exist in a variety of stages of development and with a variety of purposes and underlying philosophies. Karplus (1976) reviewed the range of approaches to modeling. He described a gradient from black box to white box models, which in his view covered a range from socioeconomic models which could be used to arouse public opinion, to models of electric circuits used for product design. While ecological and hydrological models fell relatively close together near the center of his gradient, with potential for testing theories and experimenting with control strategies, both types contain a wide array of modeling strategies. Karplus (1976) differentiated among models based on both their mathematics and their reliance on known fundamental laws. Distributed parameter models would be described using partial differential equations; lumped parameter models would use ordinary differential equations, and discrete time models would use blocks and queues. I will use these terms somewhat differently.

Two dichotomies or gradients are important for this discussion. The first gradient concerns the amount of actual physical and biological process that are incorporated in a model, as opposed to stochastic relations. An example of this dichotomy would be represented by two models of sediment erosion from a single field, the Universal Soil Loss Equation (USLE) (Wischmeier and Smith 1965) and the Chemical Runoff and Erosion from Agricultural Management Systems (CREAMS) model (Knisel 1980). The former develops indices for such factors such as vegetative cover which must be calibrated for different areas; in fact it is not very universal except that it can be applied anywhere if the calibrations have been completed (Witinok 1991). The latter is a process model which more precisely includes information on the physics of water and sediment movement.

The second gradient is that of spatial differentiation. An example of this dichotomy would be the Stanford hydrological model (Crawford and Linsley 1966), which averages all spatial variation in a drainage basin, and SHE, a grid based model which differentiates many homogeneous units within a larger heterogeneous area (Abbott *et al.* 1986*a*,*b*). Both the process-based and spatially heterogeneous models are referred to as distributed parameter models, so that the USLE is a lumped-parameter model, CREAMS and the Stanford model are distributed-parameter models, and the SHE model is distributed in both physical processes and space. It is most straightforward to distinguish between lumped- and distributed-parameter models on one hand and between spatially homogeneous and spatially heterogeneous models on the other.

Modeling the landscape

The key processes which I have discussed as interacting in the riparian landscape are hydrological, geomorphological, and ecological. It is these three areas that need to be linked with spatially explicit simulation models. Some aspects of hydrological and geomorphological models are linked already, and some ecological models include hydrological or geomorphological factors, but these models provide starting points for an as yet long and difficult task.

Hydrology

Some ecosystem developments have led to the consideration of agroecosystems and the role of riparian environments in their function. Watershed models provide the best basis on which to build a comprehensive simulation model. Bingner *et al.* (1989) provided a good review of sediment yield models for agricultural areas. Some researchers have already developed reasonably sophisticated distributed-parameter and spatially heterogeneous models (e.g. SHE (Abbott *et al.* 1986*a*,*b*);

LAVSED (Frenette and Julien 1986; Julien and Frenette 1986)). Some of these models already incorporate sediment transport and deposition and vegetative cover (e.g. the Areal Nonpoint Source Watershed Environmental Response Simulation, ANSWERS) (Beasley *et al.* 1982; cf. Rohdenburg *et al.* 1986). These models, however, primarily route water and sediment into a channel for the prediction of a hydrograph or sediment yield; they do not incorporate real geomorphological or ecological change, and, importantly, once water reaches a channel it stays there. These models do not include over-bank flooding and associated sedimentation, and so need significant modifications. Also, they are limited to relatively simple, low-gradient rivers because they do not easily move between fluvial and mass movement processes such as debris flows.

Certain hydrodynamic models do provide the input or structure for some modifications. These models originate in efforts to predict flood routing. They provide a means to model the movement of water and its kinetic force as it acts on the geomorphic structure of floodplains and as it affects the establishment, growth and mortality of plants. Hydrodynamic models have most often been used for the projection of flood conditions as an engineering problem. These models provide greatly simplified descriptions of the flow of water in and out of channels. Gee *et al.* (1990) modeled overbank flows for a landscape-scale floodplain which included topographic variation, but the complexity indicated by the addition of vegetation simply as standing pillars by Pasche & Rouve (1985) or in a simple system by Sengupta *et al.* (1986) would indicate that the inclusion of the response of falling trees and debris may best be handled stochastically. In this way the energy available for sediment transport or impact on vegetation can be generalized for a floodplain environment as a whole if not in specific locations.

Only recently have such models attained the spatial resolution or the specialization needed to concentrate specifically on the riparian environment (Lowrance and Shirohammadi 1985). ANSWERS is another event model that uses square grid cells (Beasley *et al.* 1982). It was designed with grid cells of 1–4 ha, and although originally designed for purely agricultural basins, it has been modified for use in forested catchments (Thomas and Beasley 1986). In ANSWERS, square grid cells define the drainage basin and a shadow grid is imposed on those cells that contain a channel. The channels are therefore limited to right-angle turns at the scale of the overall grid (Figure 7.1). In SHE, the channel is defined at the edge of the rectangular cells, and although the resolution can be

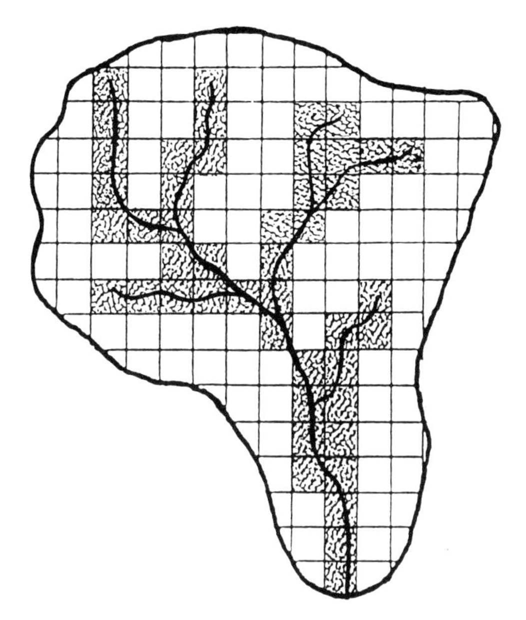

Figure 7.1 The square grid structure of ANSWERS; the stream channel is modeled using shadow cells.

increased along the river by narrowing the size of entire rows and/or columns of cells this results in over-resolution in some areas away from the river (Figure 7.2). An alternative approach for a square grid system would be to adopt a quad-tree structure in which any grid cell can be divided into four cells, and any of these cells can be further divided. Areas along the river could be given greater spatial resolution in this way, and the resolution in any area could change as the channel changed position. A second option would be to use a triangular irregular network (TIN) topographic base (Figure 7.3) (Mark 1975). The position of a channel can vary and is not limited to right angle turns. The TIN method of modeling is being increasingly adopted in hydrology and fluvial geomorphology, but it may be difficult to adapt existing ecological models to this structure (Moore *et al.* 1991). A hexagonal grid, while having useful properties for cell to cell fluxes, is less flexible in terms of adjusting the cell size.

Geomorphology

Digital terrain models provide a basis for geomorphological models. Moore *et al.* (1991) reviewed their uses for several purposes. The basic process is one in which a computer model takes an array of elevational

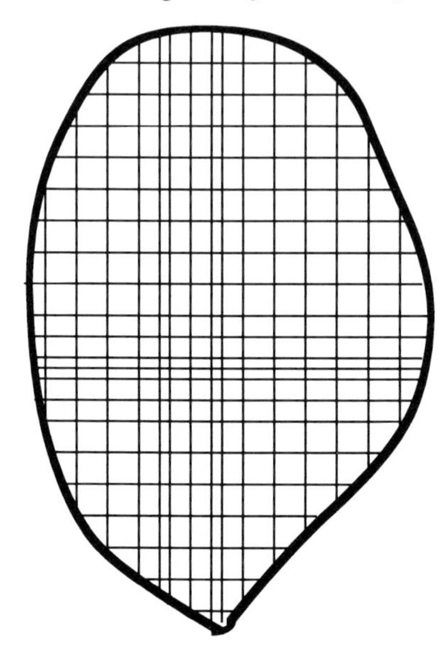

Figure 7.2 The grid structure of SHE, wherein cell dimensions are variable across the landscape so that more detail can be given to some areas, e.g. near the channel.

data and transforms it into slopes from which can be derived many topographic attributes relative to a drainage basin (e.g. Butler and Walsh 1990). These topographic maps may be based on TINs or on grids with interpolation between points. Digital terrain models are already used to provide the landform structure for use in hydrological models. The digital terrain models themselves are, however, static. Elevations, and thus slopes, drainage divides and channels, do not change.

Models of the processes of aggradation and degradation in riparian environments can be superimposed on topographic models. They are rudimentary to the extent that they do not include the effects of vegetation in the process (e.g. Odgaard 1987). The balance between erosion and deposition can shift through time for a given floodplain, but in order to understand the process in detail it will be necessary to link together several models (cf. Rhoads and Miller 1990) that can project the erosion and deposition on the floodplain, including in the channel, and that can provide a topographic basis for the hydrodynamic models and a physical basis for the third type of model.

Other geomorphological models in the form of equations for given

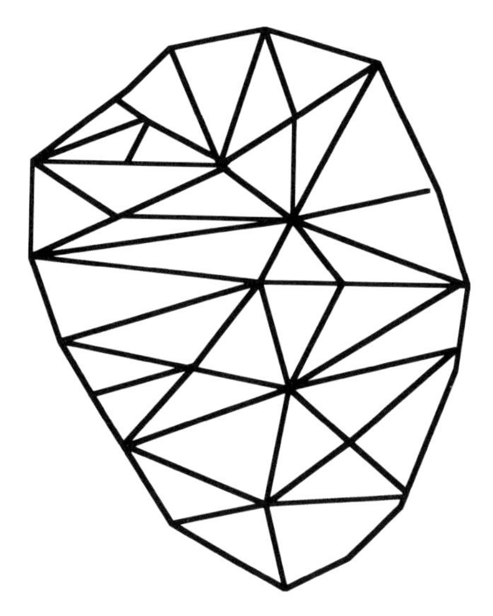

Figure 7.3 The structure of a TIN, wherein facets are defined by homogeneous slopes.

processes need to be linked to the digital terrain models interactively. Slope models, channel dynamics models, and sediment deposition models are the three of most interest. Models that portray the geomorphological development of an entire catchment, including both slope form and stream networks, would seem to be the basis for this linkage (e.g. Ahnert 1976; Willgoose *et al.* 1991*a,b,c*). These large-scale models which include both hydrological and geomorphological components are the building blocks for fully integrated landscape level models.

Ecology

The third type of model needed along with hydrological and geomorphological models is an ecological model. Unlike in hydrology and geomorphology, the value of ecological models has become a matter of intense debate. Caswell (1988) defended theoretical modeling while acknowledging its relation with empirical study. He detailed the roles of modeling in ecology which I believe are particularly appropriate to the study of riparian landscapes. His key points are that the new perspectives provided by models are important along with quantitative prediction, and that a model need not duplicate every detail of the real world, nor manipulate every important factor.

Ecological models have often emphasized either physiological or life history traits, but not both. Ecosystem models have concentrated on the relations between gross biotic structural elements, e.g. trophic levels, and cascades of energy or elements, but have tended to ignore population fluctuations. Demographic models have aimed at disclosing the relations among species, cohorts, and individuals in an environment in which the resource base, and thus fluxes of energy and matter, is generalized. The ecological models can be combined to provide a basis for interacting with the models of hydrodynamics and geomorphology. Some tradeoffs will need to be made among reality, precision, and generality in such a combined model, if for no other reason than computational excess (cf. Onstad 1988). Huston *et al.* (1988) proposed that individual-based modeling would provide a pathway to new ecological insights. Although Huston *et al.*'s (1988) implication of a new synthesis may be premature, I believe that they are on the right track. But computer simulation models relying on individual interactions must be scaled to landscape levels and must include spatial heterogeneity in order either to be useful or to have theoretical power (Goodall 1974; DeAngelis and Waterhouse 1987). The characteristics of space and place need to be incorporated hierarchically in order to understand the implications of scale, topology and history.

Models of the response of vegetation to flooding exist and are usable (e.g. Pearlstine *et al.* 1985). They do not yet interactively incorporate changes in deposition or erosion that may affect vegetation or even convert a vegetated area into channel in feedbacks. They could supply an input to hydrodynamic or geomorphological models in an iterative simulation; key factors include stem densities and sizes. The inclusion of shrub and herb layers, however, would also be necessary because of their hydraulic importance. The actual vegetation dynamics also need improvement, as noted in Chapter 4, for the relation between tree growth and depth to the water table and for the effects of nutrient inputs, as well as for general ecological relationships.

Each type of model should both respond to and provide inputs for the others. These three types of models are dynamic models of cascades of water, sediment, and information driven by a larger cascade of energy. In order to incorporate the concepts of landscape ecology and the importance of spatial structure into a process-response system, and to bring these three types of models together in a way that is programmatically feasible, a geographic information system (GIS) could be used as a base.

Space or topology in simulation models

Some spatial simulation models are in fact models which specifically produce spatial patterns rather than models of physical or biological processes which themselves produce spatial patterns. Turner (1987), for example, simulated landscape change in Georgia using a modified Markovian process involving transition probabilities in which the state of neighbors had an effect, and Pastor *et al.* (1991) used a similar approach for projecting beaver-induced changes in riparian habitat. Simulation models which do in fact include both spatial pattern and process are rare. For example, King (1991) examined spatial models in the context of scale changes and noted that the models which he examined did not include local exchanges between landscape units.

Hanson *et al.* (1989, 1990) modeled riparian forest spatial dynamics. We began with the FORFLO version (Pearlstine *et al.* 1985) of the FORET (Shugart and West 1977) forest gap dynamics model. We included rules for the dispersability of 15 riparian tree species found in Iowa, along with flooding conditions. We used a simple 3 × 40 grid of 1/12 ha plots, but only eight rows of three plots (defining one patch perpendicular to the river) were considered to be forested. We varied the distances between the patches. As a control we modeled a situation assumed by most uses of the model, what Coffin and Lauenroth (1989) called independent plots, or what we termed ubiquitous dispersal. Against this we modeled two landscapes, with greater and lesser isolation of the patches (Figure 7.4). In one paper we attempted to address the effects of including spatial pattern and process in a model when stress was included (Hanson *et al.* 1989). We changed the flow regime from high to low and vice versa and analyzed changes in species diversity. We found that with increased flooding the differences between more and less isolated patches was significant, while with decreased flow they were not. In the former case, the most isolated patches were noticeably less diverse. In analyses specifically addressing the spatial structure and process, we found that its inclusion, vs. ubiquitous dispersal, was significant. An inter-patch distance of 150 m seemed to be most important, and this is probably related to rules for dispersal of seeds by mammals in our model. We also analyzed the importance values of some of the individual species in this case. We found notable differences in the importance values of *Celtis occidentalis* (hackberry) and *Ulmus americana* (American elm). The former increases its importance value at high flow levels and especially in the most isolated sites. This result is attributable to

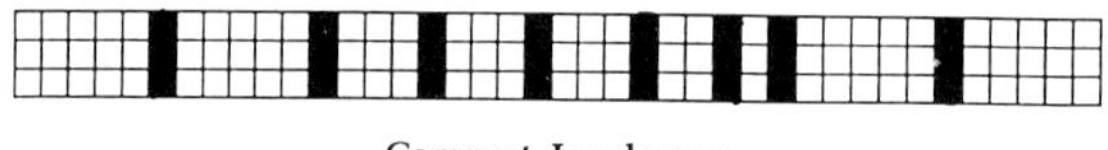

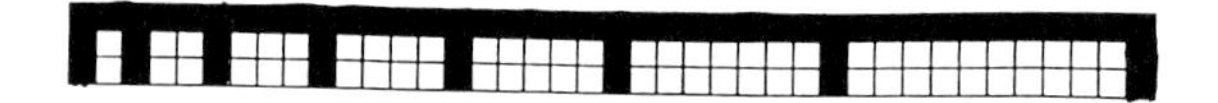

Figure 7.4 The simple grid structures used by Hanson *et al.* (1990) and Kupfer (1990*a*) to model the effects of dispersal on forest community structure in a riparian landscape.

the advantage given to *C. occidentalis* by its dispersability relative to its greatest competitors, *Carya* spp. Conversely, *U. americana* increased in abundance on the least isolated patches.

Kupfer (1990*a*) has added an edge along the discontinuous riparian corridor of the model (Figure 7.4) and found that this increased connectivity increased diversity. We are in the process of expanding the effort to a larger region, but with a focus on spatial scale and climatic change rather than on riparian landscapes (Malanson and Armstrong 1990). Kupfer (1990*b*) used the model to analyze the separate effects of climatic change and flooding during the Holocene in eastern Iowa, and Liu and Malanson (1991) found that future climatic change could interact with flooding regimes in the same region.

Geographic information systems

Geographic information systems are computer hardware and software combinations that include collection, storage, analysis, and output of spatially referenced data, i.e. data with geographic coordinates or locations (cf. Star and Estes 1990). GIS as a field of study and a means of analyzing the environment developed from computer cartography. Digital storage and representation of maps and the compilation of spatial databases burgeoned in the 1960s and 1970s. In the 1970s and 1980s, these developments led to the much more involved and complete packages that are now widely disseminated as GIS. An initial view of GIS can be thought of as a series of overlay maps, referred to as layers. Each layer contains a digital map of a feature, such as elevation or the abundance of a particular species. Basic layers can be manipulated to create additional layers, as elevation can be used to produce slope and the records of individual species can be summed to produce a community. These layers can be further manipulated to produce combinations such as the area covered by a particular combination of species on slopes of a given angle. More detailed measures of these features are also included in advanced GIS packages, such as perimeter lengths, areas of patches, and basic statistics. These methods are not without error, however, and this aspect has often been ignored (Chrisman 1984; Walsh *et al.* 1987).

GIS has been used primarily as a means to keep a spatially referenced record of environmental variables (e.g. Walsh 1985). Ashdown and Schaller (1990) provide a detailed illustration of the application of GIS technology in environmental research. They described in detail their use in cataloging and mapping different layers of information, and they presented examples of how they could be used to generate projections, using simple transfer functions, of environmental change given changing variables such as climate. Many current studies of GIS and environmental questions focus on software development or on demonstration projects. One of the uses of GIS is to link data gathered from different sources (but see error problems mentioned above). One common source is from remote sensing, especially digital imagery from satellites. Such information is now becoming common, and it has been suggested that it will prove useful in the landscape ecology of large river valleys (Girel 1986; Decamps *et al.* 1991). Hewitt (1990), Butler *et al.* (1991), and E. Muller (personal communication) have specifically mapped riparian areas using remote sensing, but improved spectral and spatial resolution are needed.

Although some GIS have the capability and purpose of providing data to a dynamic model, being updated, and entering the new data at a successive iteration, very few analyses have been made using this capability (Armstrong 1988). The use of a GIS for spatial reference in riparian forest dynamics is shown by Pearlstine *et al.* (1985). They used a GIS to record the soil type and elevations of sites and when they ran their simulation model for a given elevation type, they then mapped the distribution of the forest type using the display capability of the GIS. They did not, however, use the information in the GIS as input to the subsequent annual iterations of the model. This pattern is common in the application of GIS: they typically are not used in a dynamic interaction with temporal simulation models. Jean and Bouchard (1991) used a GIS to provide information on historical changes in riparian wetland landscapes for transition matrices and Mantel tests, and Pastor *et al.* (1991) used a GIS to provide the data to construct their Markov matrix of landscape change in an area of expanding beaver population, but they did not use it in temporal iterations which may have allowed a continuous change in the Markov components. This pattern need not be the case, as has been demonstrated, at least in concept, for the interaction of vegetation and fire (Kessell 1979).

Spatial simulation models are now the focus of research in many areas of ecology, as evidenced by numerous papers at the 1991 annual meeting of the Ecological Society of America (e.g. Levin 1991; cf. Cohen and Levin 1991). Efforts to fully integrate dynamic ecological simulations with GIS technology are under way (Malanson and Armstrong 1990; Sklar and Costanza 1991). At the most extensive level, it would be possible to program a dynamic simulation model with a GIS base which could account for the distributions of species, microclimates, material cascades, dispersal and migration, and human influences. Many of the parts of such a systems simulation are now available. It would be worthwhile to begin to explore how the connections might be made. Of ready use would be the incorporation of a dynamic phase of a GIS with the forest growth models now being used to simulate reproduction and growth of riparian forests.

A fully interactive GIS for riparian dynamics would need to include a variety of concepts that I have discussed. First, all of the riparian zone would have to be entered into a spatially referenced database. Given the state at which the dynamic models exist, a raster, or grid, format, perhaps with quad-tree capability, would be appropriate. In each grid cell the initial conditions of topography (elevation, slope and aspect), soil type, depth to water table, vegetative composition, and perhaps animal communities, would be recorded. The location of the river in the overall grid field would also be included. This field would then be the basis for simultaneously running linked models of hydrology, geomorphology and ecology.

In a given year, flood routing models would determine the magnitude and duration of flooding for a topographic field with vegetation. The output of the discharge and velocity would in part be used as input to the geomorphological model, and the hydrological model would provide information on the flooding and water table levels to the ecological model. In particular locations, such as in prograding point bars or eroding cutbanks, the location of the river itself in the spatial matrix would be changeable; the importance of this concept was noted by Phillips (1988). Over-bank flooding where elevations and vegetation are complex would mean that the distribution of water across the spatial grid would vary considerably. The subsequent fate of standing water in topographic depressions and the height of the water table would also be included.

The geomorphological model would use the discharge information from the hydrological model to calculate sediment deposition or erosion. The state of each cell, in terms of elevation, present soils, and vegetative cover, would be important to this calculation. The basic landforms could change through time, but for most simulation experiments the important processes will be in small degrees of aggradation or degradation of floodplain surfaces and both cutbank erosion and point bar deposition and growth.

The vegetation dynamics model would use the current conditions determined by the hydrological and geomorphological models to increment the establishment, growth, and mortality of the species on the grid cell. Particularly spatial characteristics would need to be considered in the ecological model. First, the accessibility of the site to propagules would be important, and the dispersability of the species at each place needs to be specified. Elements from the model SEEDFLO (Hanson *et al.* 1990) could be incorporated. Using the GIS base, the spatial structure of different layers of information would be explicitly included. Second, better definition of the physiological responses of the species is needed. Good physiological simulation models are available for only limited areas (e.g. Running and Coughlan 1988). Given the difficulty in determining parameters for many species, this effort may be delayed. In particular, the growth response to saturated soils and water table levels needs better specification than the parabolic function of Phipps (1979) and Pearlstine *et al.* (1985), and although some tree-ring evidence may be useful (Taylor 1985), the multiple factors determining growth will be difficult to separate. Moreover, while computational power and accessibility are increasing, the inclusion of both physiological and spatial processes in simulation models will consume these resources. At least in one context, i.e. the long-term effects of climatic change, observations of the Holocene record indicate that species individualistically respond by diffusion, and therefore spatial processes at an individual level may be more important than precise physiological responses.

The overall product would be a model of the riparian zone of a river in which changes in flooding will affect the topography of ridges, sloughs, abandoned channels, point bars, and cutbanks; soils will develop differentially; the conditions for species existence will vary, and a diverse array of species will exist with compositional and spatial patterns correlated to the hydrological and topographical conditions. Each of the several layers in the GIS database will be updated at each iteration of the model.

Analyses of spatial patterns

The output from a comprehensive computer simulation model of a riparian landscape will also produce spatial patterns. Turner *et al.* (1989) reviewed some means by which to analyze patterns generated by computer simulation models or by other digitized data transformations (e.g. remote sensing, digital terrain models, or other GIS layers). They proposed eight measures of landscape pattern: proportion of landscape in a category (e.g. swale dominated by *Acer saccharinum*), area and perimeter of each patch, fractal dimension of patch perimeters, edges between each pair of landcover types (e.g. *Carya*-dominated and *Acer*-dominated stands), probabilities of adjacency between land-cover types, diversity index, dominance index, and contagion index. In order to calculate many of these, however, it will be necessary to create categories. The output of the computer simulation model will be continuous. The abundance of *Carya* and *Acer* may change gradually across several grid cells, as might elevation and soil texture. These categories can be created using several statistical programs designed for such purposes. Elevation and slope groups have been created in GIS and digital terrain models; soil types can be defined; and vegetation can be classified. But the continuous nature of the changes may also be of interest. One approach will be to examine transects for changes in the gradients of particular variables, and developments in geostatistics can be applied to the output of spatially explicit models (Malanson and Armstrong 1990). A second approach will be more innovative, i.e. to use the display capabilities of the GIS to provide a basis for visual analysis. Girel (1986) has studied the problems of cartographic representation of riparian landscapes and his work will be useful for future developments, e.g. animation.

One particular area of investigation is the use of fractals for the analysis of temporal and spatial pattern. Fractal dimensions provide a measure of pattern. For example, a straight line has the dimension of 1, a complete plane has the dimension of 2, but an irregular line occupying some portion of the plane has a dimension between 1 and 2. A surface may have a fractal dimension between 2 and 3. One use of the fractal dimension may be as a dependent variable in statistical analyses or even as an independent variable when it is hypothesized that shape has an effect on a process (Rex and Malanson 1990). Milne (1991) also emphasized that fractals are directly related to scale, and so that they can be used to indicate the temporal or spatial scale at which the pattern exists at a given dimension.

Landscape models and conservation

Implications for conservation management

Nilsson (1991) has written on the issues concerning conservation of riparian landscapes. He noted five important points:

The riparian zone is connected to its landscape, and efforts at conservation must include drainage basin structure and function.

Riparian zones are rich in species.

Preservation must focus on a network of reserves.

Conservation must be oriented toward the maintenance of processes, especially hydrological regimes.

Action cannot await perfect knowledge.

The essence of all but the last point is that landscapes, not ecosystems, must be managed. Managing a landscape means managing the connectivity and flows among elements. The seasonal flooding of rivers must be maintained. The connectivity of upland and lowland ecosystems needs to be considered. The flows of energy, matter and species need to continue to reproduce a functioning landscape structure.

Managing energy flows The key element here is the flux of energy from the riparian zone to the river. Alteration of this contact, by channel modification, removal of woody debris, timber harvest, agricultural development, and other impacts all need to be structured so that the link between the terrestrial and the aquatic system is not broken.

Managing material flows The flow most altered is that of water. Reservoirs and diversions have all but eliminated purely natural flows from most riparian environments. Conservation will require some modification of flow regimes by including nature as a value in addition to water supply, hydroelectric generation, flood control and recreation.

Flows of sediment also need to be managed. Hoping that riparian zones will continue to store sediment and keep it from the river ignores its impact on the riparian zone itself. Hunt (1988) recommended maintenance of in-stream flow as the single most important problem for riparian environmental quality.

Managing species flows Nilsson's (1991) idea of a network of reserves needs to be developed for specific species and landscapes. The function of riparian zones as corridors between larger reserves needs to be used to advantage.

While it is true that action should not await perfect knowledge, and that inaction is likely to be the worst alternative, additional knowledge is needed in real conservation and management efforts. It is needed for two reasons: to combat forces aimed at environmental destruction and to guide best use of limited conservation and management resources (e.g. Pinay *et al.* 1988).

Much environmental destruction is caused by economic forces driven by greed and/or ignorance, along with associated poverty, necessity, and accident. The economic development of rivers and riparian areas is often for the short-term profit of a few individuals, at a long-term loss to society in general. In order to combat these forces, the values of riparian areas to society must be understood and communicated, and because the values will be challenged by the forces of economic development, they must be scientifically supportable. This valuation requires better knowledge of how the riparian zone works in the landscape context. Past regulatory efforts have addressed only local and piecemeal impacts rather than cumulative effects, and the full range of destruction has not been known. Cooperation, rather than antagonism, between managing groups is needed (Hunt 1988; Knopf *et al.* 1988).

One approach to resolving conflicts is Multiobjective River Corridor management (MORC) (Melanson 1991). This process will not end conflict, but does provide a basis for competing interests to express their desires and, perhaps, to understand the desires of others. MORC is a concept that specifically takes into account the fact that different goals of river corridor management will produce conflicts. While MORC is addressed to 'planners', it specifically calls for public participation. It also recognizes the important differences in local geography and long-term concerns. The process seems overly optimistic, however, in that it does not seem to recognize that some conflicts may be irreconcilable, i.e. that alternative individual goals which may be desirable in a general sense may be mutually incompatible; that those who choose to participate may not represent the public and that relations of power will affect participation; and that scientific information can be manipulated or used with different success by different interest groups (Hollander 1991). None the less, MORC is a starting point.

Conservationists must deal with limited resources. It will not be possible to restore entire large drainage basins that are already populated and impacted by human activities, and a choice of areas within basins on which to focus conservation and preservation efforts must be made. Restoration projects, while adding to our knowledge of riparian processes, are costly and uncertain (Cairns *et al.* 1979; Anderson and Ohmart 1985; Clewell and Lea 1990; US Army Corps of Engineers 1991; Youngblood and Zasada 1991). Financial and human resources need to be allocated to the areas which will have the greatest net positive effect on the landscape. The allocation requires a good understanding of the riparian processes at the landscape scale.

Although we cannot await perfect knowledge, more and better information is still needed. A notable need is that conservation management must be at a landscape scale, while ecological research has been at an ecosystem scale. Much of the information from ecosystem-level studies is difficult to use in management because of the scale problem. The landscape ecology of riparian conservation needs work. Although certain functions have been identified as important, their link to structure in landscape ecology is unclear, and the conservation management of landscape structure provides a starting point with a firm foundation. One of the areas where research for conservation is likely to be important and fruitful is on the issue of biodiversity in riparian landscapes.

Landscape conservation and biodiversity

Landscape ecology recognizes that disturbance may be an integral part of the natural processes of an area and may serve to help maintain biodiversity. Basic theoretical statements (Forman and Godron 1986) and symposia directed at this topic (Turner 1987) have not been clear about the distinction between the effects of disturbance on alpha diversity, i.e. the number of species on a particular site, and gamma diversity, i.e. the number of species in a region. The frequency and intensity of disturbance at any one place should be distinguished from the landscape-level effects of disturbance. The intermediate disturbance hypothesis was proposed by Connell (1978) for the processes allowing coexistence at a single site, i.e. processes maintaining alpha diversity. He hypothesized that species that were adapted to either high or low levels of disturbance would coexist, but did not consider that species adapted specifically to the intermediate level could exclude those from either extreme (Malanson 1987). In addition to the intermediate disturbance hypothesis a corollary variable disturbance hypothesis seems necessary (Malanson 1992). Using models, the frequency of disturbance and the variability of the frequency have been shown to allow the coexistence of species with different niches along the frequency gradient (e.g. Fagerstrom and Agren 1979). The variability of the disturbance itself can promote coexistence and eliminates the problem of a superior competitor adapted specifically to the intermediate level of disturbance.

The extension of this hypothesis to propose that disturbances within a landscape that in sum are intermediate in frequency and intensity is, however, natural. Landscapes can be considered to be shifting mosaics of species and/or communities (Whittaker and Levin 1977). Disturbance is the primary element driving the spatially progressive nature of this dynamic process. Disturbances create multiple landscape elements and allow the coexistence of species, not necessarily at the same place and time, but within the same landscape, and thus increase gamma diversity.

Salo *et al.* (1986) argued that an increase in biodiversity in the Amazon, due to disturbance, is a landscape-level phenomenon, i.e. fluvial dynamics create different 'tesserae' or subunits of landscape elements (*sensu* Forman and Godron 1986) and increase gamma, or regional, diversity. Salo (1990) has since phrased the problem in more distinct landscape ecology terms, noting that exogenous forces create the multitude of landforms that make up the extensive riparian forest of the Amazon. Disturbance processes have created several tesserae in riparian landscape elements. Where channel dynamics and flooding create disturbances, the species richness may be greater because a greater variety of environments and of successional seres can coexist within the landscape element. Distinct plant communities exist on point bars, in abandoned channels, at cutbank edges, and in extensive floodplain interiors. The biodiversity of these several communities is maintained by active fluvial processes. Differences in both life history and physiological traits among species separates them among tesserae. The disturbance expands the range of environmental gradients available. In contrast to some other environments, where the disturbance primarily allows ruderal or fugitive species to occupy transient sites, in riparian areas the expanded gradients include very long-term changes in site conditions, such as in abandoned channels, as well as transient sites such as point bars and channel bars.

The spatially transgressive nature of an active river also imposes nonequilibrium conditions in riparian vegetation (Kupfer and Malanson 1992*a*). The topological progressions of a meander creates a particular local disequilibrium which is complemented by point bar development (cf. Kalliola and Puhakka 1988) and leads to the moderating influence of spatial extent and heterogeneity proposed by DeAngelis and Waterhouse (1987) for landscapes. At a site, intensity may be thought of as the flood power and sedimentation that causes a given amount of damage to the biota, but at the landscape level, the relations between intensity, frequency and spatial scale is emphasized because the numbers, sizes and locations of landforms, such as point bars, are variable and depend on the disturbance events, and the numbers, sizes, and isolation of sites can affect their response to fluvial disturbances (Malanson 1982).

Studies in riparian areas have not demonstrated the long-term coexistence of species adapted to different degrees of disturbance on a single site which experiences an intermediate level of disturbance. The spatially progressive nature of fluvial erosion and deposition defines the problem for most riparian studies. Single locations do not experience a disturbance regime that has a constant mean and variance. Instead, as on point bars, the site experiences progressively less flooding and sedimentation. Where we see coexistence of species adapted to higher and lower levels of disturbance in this environment is not at a site that is maintained by continuing intermediate disturbance levels, but at a site that is intermediate in its transition from high to low disturbance levels.

The riparian landscape, by providing several instances of spatial transformations, provides an excellent area for the future exploration of the intermediate and variable disturbance hypotheses relative to gamma diversity at the landscape scale. The controlling factor is the way in which the landscape reproduces itself. It is not simply a matter of species reproduction in a single place, but it involves the creation and destruction of landforms, the topological dynamics of hydrology, and the dispersal, establishment, growth and death of organisms; all of these factors are processes of disturbance in and of themselves, as well as structures that respond to disturbances. The number of species that can coexist in the landscape is determined by its reproduction, and in turn the species affect the reproduction through their interactions and their effects on the physical environment. Diversity is a variable of interest in simulation models, and spatially explicit models at the landscape scale will allow a better distinction between alpha and gamma diversity.

Conclusions

A general qualitative picture of the processes of riparian environments does emerge: erosion and deposition interact with forest dynamics to control the movement of energy and material through the system. This concept is most apparent in the extensive floodplains of alluvial rivers where both erosion and deposition occur and one can succeed the other at any given place. Erosional channels are abandoned and become sites for deposition. Ecological succession occurs and the site concentrates energy and material. Disturbances and stresses, such as fires, droughts, and livestock grazing, alter the response. Renewed erosion is resisted but is inevitable. The material deposited between erosion events is then transferred downstream. A rich ecological process is played out on this shifting substrate.

There are no permanent sinks for energy or material. The riparian environment imposes a pulse on what could otherwise be a continuous flow, but it also modifies the pulse of exogenous inputs. Flood flows are held back, but nutrients, sediment and information are released in pulses which are longer in duration and reduced in intensity compared with the function of a river without such a system. The floodplain and the river are moving in space. The channel hydraulics are the essential driving force of this model. This spatial dynamic controls the internal structure, through the transverse gradient of distance from the channel, but is modified by the temporal gradient of time since the site was itself a channel. The vegetation and wildlife respond to and in turn control this dynamic. Dense seedling establishment, massive amounts of woody debris, and the structures of beaver are part of this interaction. The longitudinal gradient along the river course does not come into play in a regionally stable environment. If, however, the climate or major aspects of the river regime change, then the linear form of the river may serve to hinder or promote the spatial adjustment of species; this role is yet unknown.

Landscape ecology provides a useable and flexible perspective on the problems of nature reserves and ecological management in general. The approach is capable of directly addressing the concerns of decision-makers while developing fundamental insights into the processes that maintain diversity and populations of particular species. Riparian landscapes are important loci for management because of their unique values, their physical and biological characteristics, and their spatial configuration. Environmental management and science can develop together in this landscape, and this union should prove very productive in the near future.

References

Abbott, M.B., Bathurst, J.C., Cunge, J.A., O'Connell, P.E. and Rasmussen, J. 1986*a*. An introduction to the European hydrological system – Systeme Hydrologique Europeen, 'SHE', **1**: History and philosophy of a physically-based, distributed modelling system. *Journal of Hydrology* **87**: 45–59.

Abbott, M.B., Bathurst, J.C., Cunge, J.A., O'Connell, P.E. and Rasmussen, J. 1986*b*. An introduction to the European hydrological system – Systeme Hydrologique Europeen, 'SHE', **2**: Structure of a physically-based, distributed modelling system. *Journal of Hydrology* **87**: 61–77.

Abell, D.L. (ed.) 1989. *Proceedings of the California Riparian Systems Conference*. USA Forest Service General Technical Report PSW-110.

Abernethy, Y and Turner, R.E. 1987. USA forested wetlands: 1940–1980. *BioScience* **37**: 721–7.

Abler, R., Adams, J.S. and Gould, P. 1971. *Spatial Organization*. Prentice-Hall, Englewood Cliffs, NJ.

Abrams, M.D. 1985. Age-diameter relationships of *Quercus* species in relation to edaphic factors in gallery forests in northeast Kansas. *Forest Ecology and Management* **13**: 181–93.

Abrams, M.D. 1986. Historical development of gallery forests in northeast Kansas. *Vegetatio* **65**: 29–37.

Adams, D.E. and Anderson, R.C. 1980. Species response to a moisture gradient in central Illinois forests. *American Journal of Botany* **67**: 381–92.

Adis, J., Furch, K. and Irmler, U. 1979. Litter production of a Central-Amazonian black water inundation forest. *Tropical Ecology* **20**: 236–45.

Ahnert, F. 1976. Brief description of a comprehensive three-dimensional process–response model for landform development. *Zeitschrift für Geomorphologie*, Supplementband **25**: 29–49.

Aikman, J.M. 1930. Secondary plant succession on Muscatine Island, Iowa. *Ecology* **11**: 577–88.

Aikman, J.M. and Gilly, C.L. 1948. A comparison of the forest floras along the Des Moines and Missouri Rivers. *Proceedings of the Iowa Academy of Science* **55**: 53–73.

Ain, G., Gilot, B., Neuburger, M.C., Pautou, G., Tetart, and Thomas, J. 1973. *Étude écologique des anciens lits du Rhône entre le confluent du Guiers et le confluent de l'Ain.*

Laboratoire de Biologie Végétale, Université Scientifique et Medicale de Grenoble, Grenoble.

Aleksiuk, M. 1970. The seasonal food regime of arctic beavers. *Ecology* **51**: 264–70.

Alexander, C.S. and Prior, J.C. 1971. Holocene sedimentation rates in overbank deposits in the Black Bottom of the lower Ohio River, southern Illinois. *American Journal of Science* **270**: 361–72.

Allen, J.R.L. 1965. A review of the origin and characteristics of recent alluvial sediments. *Sedimentology* **5**: 89–191.

Allred, M. 1980. A re-emphasis on the value of the beaver in natural resource conservation. *Journal of the Idaho Academy of Science* **16**: 3–10.

Ambus, P. and Lowrance, R. 1991. Comparison of denitrification in two riparian soils. *Soil Science Society of America Journal* **55**: 994–7.

Amoros, C. and Jacquet, C. 1986. Évolution des anciens méandres: diagnostic base sûr les restes de Cladoceres (Crustaces) conservés dans les sédiments. *Documents de Cartographie Ecologique* **29**: 135–46.

Amoros, C., Bravard, J.P., Pautou, G., Reygrobellet, J.L. and Roux, A.L. 1986. Synthèses, prévisions et gestion écologique. *Documents de Cartographie Ecologique* **29**: 147–60.

Amoros, C., Roux, A.L., Reygrobellet, J.L., Bravard, J.P. and Pautou, G. 1987*a*. A method for applied ecological studies of fluvial hydrosystems. *Regulated Rivers: Research and Management* **1**: 17–36.

Amoros, C., Rostan, J.-C., Pautou, G. and Bravard, J.P. 1987*b*. The reversible process concept applied to the environmental management of large river systems. *Environmental Management* **11**: 607–17.

Anderson, B.W. and Ohmart, R.D. 1985. Riparian revegetation as a mitigating process in stream and river restoration. In J.A. Gore (ed.) *The Restoration of Rivers and Streams*, pp. 41–79, Butterworth, Boston.

Anderson, E. 1947. Missouri gravel bars. *Missouri Botanical Garden Bulletin* **35**: 166.

Anderson, M.G. and Rogers, C.C.M. 1987. Catchment scale distributed hydrological models: a discussion of research directions. *Progress in Physical Geography* **11**: 29–51.

Anderson, M.T. 1985. Riparian management of coastal Pacific ecosystems. In R.R. Johnson, C.D. Ziebell, D.R. Patton, P.F. Ffolliott and R.H. Hamre (eds.) *Riparian Ecosystems and Their Management: Reconciling Conflicting Uses*. US Forest Service General Technical Report RM-120, pp. 364–8.

Andrews, P., Groves, C.P. and Horne, J.F.M. 1975. Ecology of the lower Tana River flood plain (Kenya). *Journal of the East Africa Natural History Society and National Museum* **151**: 1–31.

Andrus, C.W., Long, B.A. and Froehlich, H.A. 1988. Woody debris and its contribution to pool formation in a coastal stream 50 years after logging. *Canadian Journal of Fisheries and Aquatic Sciences* **45**: 2080–6.

Ansari, T.A. 1961. Riverain forests of Sind – hope or despair. *Empire Forestry Review* **40**: 228–233.

Apple, L.L. 1985. Riparian habitat restoration and beavers. In R.R. Johnson, C.D. Ziebell, D.R. Patton, P.F. Ffolliott and R.H. Hamre (eds.) *Riparian Ecosystems and Their Management: Reconciling Conflicting Uses*. US Forest Service General Technical Report RM-120, pp. 489–90.

Arcement, G.J. and Schneider, V.R. 1989. Guide for selecting Manning's roughness coefficients for natural channels and floodplains. US Geological Survey Water-Supply Paper 2339.

Armstrong, M.P. 1988. Temporality in spatial databases. In *Proceedings GIS–LIS '88* (American Congress on Surveying and Mapping, Falls Church, Virginia), pp. 880–9.

Ashdown, M. and Schaller, J. 1990. *Geographic Information Systems and Their Application in MAB-Projects, Ecosystem Research and Environmental Monitoring*. German National Committee for the UNESCO Programme 'Man and the Biosphere' (MAB), Bonn.

Athearn, F.J. 1988. *Habitat in the Past: Historical Perspectives of Riparian Zones on the White River*. Cultural Resource Series No. 23, Bureau of Land Management, Colorado State Office, Denver.

Atkin, D. 1980. The Age Structure and Origins of Trees in a New Jersey Floodplain forest. Ph.D. dissertation, Princeton University.

Attwell, I.G. 1970. Some effects of Lake Kariba on the ecology of a floodplain of the mid-Zambezi valley of Rhodesia. *Biological Conservation* **2**: 189–96.

Austin, M.P. 1987. Models for the analysis of species' response to environmental gradients. *Vegetatio* **69**: 33–45.

Austin, T.A., Riddle, W.F. and Landers, R.Q. 1979. Mathematical modeling of vegetative impacts from fluctuating flood pools. *Water Resources Bulletin* **15**: 1265–80.

Bacon, P.R. 1990. Ecology and management of swamp forests in the Guianas and Caribbean region. In: A.E. Lugo, M.Brinson, and S. Brown (eds.) *Forested Wetlands* (*Ecosystems of the World* 15), pp. 213–50. Elsevier, Amsterdam.

Bagi, I. 1987. The vegetation map of the Kisapaj UNESCO Biosphere Reserve Core Area, Kiskunsag National Park, Hungary. *Acta Biologica Szeged.* **33**: 63–74.

Bailey, R.G. 1976. *Ecoregions of the Unites States* (map). USDA Forest Service Intermountain Region, Ogden, Utah.

Bailey, R.G. 1980. *Descriptions of the Ecoregions of the United States*. USDA Forest Service Miscellaneous Publication No. 1391, Washington.

Bailey, R.G. 1983. Delineation of ecosystem regions. *Environmental Management* **7**: 365–73.

Bailey, R.G. 1989. Explanatory supplement to ecoregions map of the continents. *Environmental Conservation* **16**: 307–309 (and map).

Baird, L. and Anderson, M.G. 1992. Ungauged catchment modelling. I. Assessment of flood plain flow model enhancements. *Catena* **19**: 17–31.

Baker, R.G., Chumbley, C.A., Witinok, P.M. and Kim, H.K. 1990. Holocene vegetational changes in eastern Iowa. *Journal of the Iowa Academy of Science* **97**: 167–77.

Baker, R.G., Horton, D.G., Kim, H.K., Sullivan, A.E., Roosa, D.M., Witinok, P.M. and Pusateri, W.P. 1987. Late Holocene paleoecology of southeastern Iowa: development of riparian vegetation at Nichols Marsh. *Proceedings of the Iowa Academy of Science* **94**: 51–70.

Baker, W.L. 1988. Size-class structure of contiguous riparian woodlands along a Rocky Mountain river. *Physical Geography* **9**: 1–14.

Baker, W.L. 1989. Macro- and micro-scale influences on riparian vegetation in western Colorado. *Annals of the Association of American Geographers* **79**: 65–78.

Baker, W.L. 1990*a*. Climatic and hydrologic effects on the regeneration of *Populus angustifolia* James along the Animas River, Colorado. *Journal of Biogeography* **17**: 59–73.

Baker, W.L. 1990*b*. Species richness in Colorado riparian vegetation. *Journal of Vegetation Science* **1**: 119–24.

Bakker, J.P., Brouwer, C., van den Hof, L. and Jansen, A. 1987. Vegetational succession, management and hydrology in a brookland (The Netherlands). *Acta Botanica Neerlandica* **36**: 39–58.

Balslev, H., Luteyn, J., Ollgaard, B. and Holm-Nielsen, L.B. 1987. Composition and structure of adjacent unflooded and floodplain forest in Amazonian Ecuador. *Opera Botanica* **92**: 37–57.

Band, L.E. and Wood, E.F. 1988. Strategies for large-scale, distributed hydrologic simulation. *Applied Mathematics and Computation* **27**: 23–37.

Barnes, W.J. 1978. The distribution of floodplain herbs as influenced by annual flood elevation. *Wisconsin Academy of Sciences, Arts and Letters Transactions* **66**: 254–66.

Barnes, W.J. 1985. Population dynamics of woody plants on a river island. *Canadian Journal of Botany* **63**: 647–55.

Barnes, W.J. and Dibble, E. 1988. The effects of beaver in riverbank forest succession. *Canadian Journal of Botany* **66**: 40–4.

Bartel-Ortiz, L.M. and David, M.B. 1988. Sulfur constituents and transformations in upland and floodplain forest soils. *Canadian Journal of Forest Research* **18**: 1106–12.

Barton, D.R., Taylor, W.D. and Biette, R.M. 1985. Dimensions of riparian buffer strips required to maintain trout habitat in southern Ontario streams. *North American Journal of Fisheries Management* **5**: 364–78.

Baum, B.R. 1967. Introduced and naturalized tamarisks in the United States and Canada. *Baileya* **15**: 19–25.

Baumann, R.H., Day, J.W. and Miller, C.A. 1984. Mississippi deltaic wetland survival: sedimentation versus coastal submergence. *Science* **224**: 1093–5.

Beasley, D.B., Huggins, L.F. and Monke, E.J. 1982. Modeling sediment yields from agricultural watersheds. *Journal of Soil and Water Conservation* **37**: 113–17.

Bedinger, M.S. 1971. Forest species as indicators of flooding in the lower White River valley, Arkansas. USA Geological Survey Professional Paper 750-C.

Bedinger, M.S. 1978. Relation between forest species and flooding. In *Wetland Functions and Values: The State of Our Understanding*, pp. 427–35. American Water Resources Association, Baltimore.

Bedinger, M.S. 1979. Forests and flooding with special reference to the White River and Ouachita River Basins, Arkansas. US Geological Survey Water Resources Investigations Open File Report 79–68.

Bedinger, M.S. 1981. Hydrology of bottomland hardwood forests of the Mississippi embayment. In J.R. Clark and J. Benforado (eds.) *Wetlands of Bottomland Hardwood Forests*, pp. 161–78. Elsevier, New York.

Beerling, D.J. 1991. The effect of riparian land use on the occurrence and abundance of Japanese knotweed *Reynoutria japonica* on selected rivers in South Wales. *Biological Conservation* **55**: 329–37.

Begin, Y. and Lavoie, J. 1988. Dynamique d'une bordure forestière et variations récentes du niveau du fleuve Saint-Laurent. *Canadian Journal of Botany* **66**: 1905–13.

Bell, D.T. 1974*a*. Tree stratum composition and distribution in the streamside forest. *American Midland Naturalist* **92**: 35–46.

Bell, D.T. 1974*b*. Studies on the ecology of a streamside forest: composition and distribution of vegetation beneath the tree canopy. *Bulletin of the Torrey Botanical Club* **101**: 14–20.

Bell, D.T. 1980. Gradient trends in the streamside forest of central Illinois. *Bulletin of the Torrey Botanical Club* **107**: 172–80.

Bell, D.T. and del Moral, R. 1977. Vegetation gradients in the streamside forest of Hickory Creek, Will County, Illinois. *Bulletin of the Torrey Botanical Club* **104**: 127–35.

Bell, D.T. and Johnson, F.L. 1974. Ground-water level in the flood plain and adjacent uplands of the Sangamon River. *Transactions of the Illinois State Academy of Science* **67**: 376–83.

Bell, D.T. and Sipp, S.K. 1975. The litter stratum in the streamside forest ecosystem. *Oikos* **26**: 391–7.

Bell, D.T., Johnson, F.L. and Gilmore, A.R. 1978. Dynamics of litter fall, decomposition, and incorporation in the streamside forest ecosystem. *Oikos* **30**: 76–82.

Bellah, R.G. and Hulbert, L.C. 1974. Forest succession on the Republican River in Clay County, Kansas. *Southwestern Naturalist* **19**: 155–66.

Beltman, B. and Grootjans, A.P. 1986. Distribution of nutrient poor plant communities in relation to the groundwater regime and nutrient availability. In *Water Management in Relation to Nature, Forestry and Landscape Management*, pp. 59–79. TNO Committee on Hydrological Research, Proceedings and Information No. 34, The Hague.

Beschta, R.L. 1979. Debris removal and its effect on sedimentation in an Oregon Coast Range stream. *Northwest Science* **53**: 71–7.

Bettis, E.A., Baker, R.G., Nations, B.K. and Benn, D.W. 1990. Early Holocene pecan, *Carya illinoensis*, in the Mississippi River valley near Muscatine, Iowa. *Quaternary Research* **33**: 102–7.

Bhowmik, N.G. and Demissie, M. 1986. Momence Wetland: influence on sediment and water. In P.G. Sly (ed.) *Sediments and Water Interactions*, pp. 71–7. Springer-Verlag, New York.

Bhowmik, N.G. and Demissie, M. 1989. Sedimentation in the Illinois River valley and backwater lakes. *Journal of Hydrology* **105**: 187–95.

Bhowmik, N.G., Adams, J.R. and Demissie, M. 1988. Sedimentation of four reaches of the Mississippi and Illinois Rivers. Sediment Budgets. International Association of Hydrological Sciences (IAHS) Publication no. 174, pp. 11–19.

Bilby, R.E. 1981. Role of organic debris dams in regulating the export of dissolved and particulate matter from a forested watershed. *Ecology* **62**: 1234–43.

Bilby, R.E. 1984. Post-logging removal of woody debris affects channel stability. *Journal of Forestry* **82**: 609–13.

Bilby, R.E. and Likens, G.E. 1980. Importance of organic debris dams in the structure and function of stream ecosystems. *Ecology* **61**: 1107–13.

Bingner, R.L., Murphree, C.E. and Mutchler, C.K. 1989. Comparison of sediment yield models on watersheds in Mississippi. *Transactions of the ASAE* **32**: 529–34.

Birdsell, J.B. 1950. Some implications of the genetical concept of race in terms of spatial analysis. *Cold Spring Harbor Symposium in Quantitative Biology* **15**: 259–311.

Bliss, L.C. and Cantlon, J.E. 1957. Succession on river alluvium in northern Alaska. *American Midland Naturalist* 58, 452–69.

Blom, C.W.P.M., Bogemann, G.M., Laan, P., van der Sman, A.J.M., van de Steeg, H.M. and Voesenek, L.A.C.J. 1990. Adaptations to flooding in plants from river areas. *Aquatic Botany* **38**: 29–47.

Boedeltje, G. and Bakker, J.P. 1980. Vegetation, soil, hydrology and management in a Drenthian brookland (The Netherlands). *Acta Botanica Neerlandica* **29**: 509–22.

Boldt, C.E., Uresk, D.W. and Severson, K.E. 1979. Riparian woodland enclaves in the prairie draws of the northern High Plains: a look at problems, a search for solutions. *Great Plains Agricultural Council* **91**: 31–32.

Boles, P.H. and Dick-Peddie, W.A. 1983. Woody riparian vegetation patterns on a segment of the Mimbres River in southwestern New Mexico. *Southwestern Naturalist* **28**: 81–7.

Bormann, F.H. and Likens G.E. 1979. *Pattern and Process in a Forested Ecosystem.* Springer-Verlag, New York.

Bournaud, M. and Amoros, C. 1984. Des indicateurs biologiques aux descripteurs de fonctionnement: quelques exemples dans un système fluvial. *Bulletin d'Ecologie* **15**: 57–66.

Bowers, L.J., Gosselink, J.G., Patrick, W.H. and Choong, E.T. 1985. Influence of climatic trends on wetland studies in the eastern United States which utilize tree ring data. *Wetlands* **5**: 191–200.

Bowman, D.M.J.S. and McDonough, L. 1991. Tree species distribution across a seasonally flooded elevation gradient in the Australian monsoon tropics. *Journal of Biogeography* **18**: 203–12.

Bowser, C.W. 1960. The phreatophyte problem. *Proceedings of the Arizona Watershed Symposium* **4**: 17–18.

Bradford, J.M. and Piest, R.F. 1977. Gully wall stability in loess-derived alluvium. *Soil Science Society of America Journal* **41**: 115–22.

Bradley, C.E. and Smith, D.G. 1986. Plains cottonwood recruitment and survival on a prairie meandering river floodplain, Milk River, southern Alberta and northern Montana. *Canadian Journal of Botany* **64**: 1433–42.

Braun-Blanquet, J. 1915. Les Cevennes meridionales (Massif de l'Aigoual). *Études sûr la végétation mediterranéenne.* Société Physique Naturelle, Geneva.

Braun-Blanquet, J. 1951. *Pflanzensoziologie.* Springer-Verlag, Vienna.

Bravard, J.P. 1982. A propos de quelques formes fluviales de la vallée du haut-Rhône français. *Revue de Géographie de Lyon* **1982/1**: 39–48.

Bravard, J.P. 1983. Les sédiments fins des plaines d'inondation dans la vallée du Haut-Rhône. *Revue de Géographie Alpine* **71**: 363–79.

Bravard, J.P. 1986. La basse vallée d l'Ain: dynamique fluviale appliquée a l'écologie. *Documents de Cartographie Ecologique* **29**: 17–43.

Bravard, J.P. 1987. *Le Rhône du Leman a Lyon.* La Manufacture, Paris.

Bravard, J.P. 1989. La métamorphose des rivières des Alpes françaises a la fin du moyen-age et a l'epoque moderne. *Bulletin de la Société Géographique de Liege* **25**: 145–57.

Bravard, J.P. 1990. Observations nouvelles sur la dynamique fluviale et l'alluvionnement de la Saone à l'Holocène, entre Villefranche et Anse (Rhone). *Revue Géographique de l'Est* **1990** (1): 57–76.

Bravard, J.P., Amoros, C. and Pautou, G. 1986. Impact of civil engineering works on the successions of communities in a fluvial system. *Oikos* **47**: 92–111.

Bray, D.I. 1987. A study of channel changes in a reach of the North Nashwaaksis stream, New Brunswick, Canada. *Earth Surface Processes and Landforms* **12**: 151–65.

Brayton, D.S. 1984. The beaver and the stream. *Journal of Soil and Water Conservation* **39**: 108–9.

Bren, L.J. 1987. The duration of inundation in a flooding river red gum forest. *Australian Forestry Research* **17**: 191–202.

Bren, L.J. 1988. Flooding characteristics of a riparian red gum forest. *Australian Forestry* **51**: 57–62.

Bren, L.J. and Gibbs, N.L. 1986: Relationships between flood frequency, vegetation and topography in a river red gum forest. *Australian Forest Research* **16**: 356–70.

Bren, L.J., O'Neill, I.C. and Gibbs, N.L. 1988. Use of map analysis to elucidate flooding in an Australian riparian river red gum forest. *Water Resources Research* **24**: 1152–62.

Bridgewater, P.B. 1980. Fringing forests of the Swan River, Western Australia: past, present and future. *La Vegetation des Forets Alluviales. Colloques Phytosociologiques* **9**: 325–42.

Brinson, M.M. 1977. Decomposition and nutrient exchange in an alluvial swamp forest. *Ecology* **58**: 601–9.

Brinson, M.M. 1985. Management potential for nutrient removal in forested wetlands. In P.J. Godfrey, E.R. Kaynor, S. Pelczarski and J. Benforado (eds.) *Ecological Considerations of Wetland Treatment of Municipal Wastewater*, pp. 405–14. Van Nostrand Reinhold Co., New York.

Brinson, M.M. 1990. Riverine forests. In: A.E. Lugo, M.Brinson and S. Brown (eds.) *Forested Wetlands. Ecosystems of the World 15*, pp. 87–141. Elsevier, Amsterdam.

Brinson, M.M., Bradshaw, H.D., Holmes, R.N. and Elkins, J.B. 1980. Litterfall, stemflow, and throughfall nutrient fluxes in an alluvial swamp forest. *Ecology* **61**: 827–35.

Brinson, M.M., Bradshaw, H.D. and Holmes, R.N. 1983. Significance of floodplain sediments in nutrient exchange between a stream and its floodplain. In T.D. Fontaine and S.M. Bartell (eds.) *Dynamics of Lotic Ecosystems*, pp. 199–221. Ann Arbor Science, Ann Arbor, Michigan.

Brinson, M.M., Swift, B.L., Plantico, R.C. and Barclay, J.S. 1981. *Riparian Ecosystems: Their Ecology and Status.* USA Fish and Wildlife Service OBS-81/17, Washington.

Broadfoot, W.M. and Williston, H.L. 1973. Flooding effects on southern forests. *Journal of Forestry* **71**: 584–7.

Brothers, T.S. 1984. Historical vegetation change in the Owens River riparian woodland. In R.E. Warner and K.M. Hendrix (eds.) *California Riparian Systems*, pp. 75–84. University of California Press, Berkeley.

Brown, C.E. 1988. Physicochemical Characteristics of Nine First Order Streams Under Three Riparian Management Regimes in East Texas. M.S. thesis, Stephen F. Austin State University, Nacogdoches, Texas.

Brown, D.E. and Lowe, C.H. 1974. The Arizona system for natural and potential vegetation – illustrated summary through the fifth digit for the North American Southwest. *Journal of the Arizona Academy of Sciences* **9** (Supplement): 1–28.

Brown, K.R., Zobel, D.B. and Zasada, J.C. 1988. Seed dispersal, seedling emergence, and early survival of *Larix laricina* (DuRoi) K. Koch in the Tanana Valley, Alaska. *Canadian Journal of Forest Research* **18**: 306–14.

Brown, S. 1981. A comparison of the structure, primary productivity and transpiration of cypress ecosystems in Florida. *Ecological Monographs* **51**: 403–27.

Brown, S. and Peterson, D.L. 1983. Structural characteristics and biomass production of two Illinois bottomland forests. *American Midland Naturalist* **110**: 107–17.

Brown, S., Brinson, M.M. and Lugo, A.E. 1979. Structure and function of riparian wetlands. In R.R. Johnson and J.F. McCormick (eds.) *Strategies for Protection and Management of Floodplain Wetlands and Other Riparian Ecosystems*. US Forest Service General Technical Report WO-12, pp. 17–31.

Brown, T.C. and Daniel, T.C. 1991. Landscape aesthetics of riparian environments: relationship of flow quantity to scenic quality along a Wild and Scenic River. *Water Resources Research* **27**: 1787–95.

Bruenig, E.F. 1990. Oligotrophic forested wetlands in Borneo. In: A.E. Lugo, M. Brinson and S. Brown (eds.) *Forested Wetlands. Ecosystems of the World 15*, pp. 299–334. Elsevier, Amsterdam.

Bryant, M. 1980. Evolution of large, organic debris after timber harvest: Maybeso Creek, 1949–1978. USA Forest Service General Technical Report PNW-101.

Buchholz, K. 1981. Effects of minor drainages on woody species distributions in a successional floodplain forest. *Canadian Journal of Forest Research* **11**: 671–6.

Buckley, R. 1981. Alien plants in central Australia. *Botanical Journal of the Linnean Society* **82**: 369–79.

Buell, M.F. and Wistendahl, W.A. 1955. Flood plain forests of the Raritan River. *Bulletin of the Torrey Botanical Club* **82**: 463–72.

Burdick, D.M., Cushman, D., Hamilton, R. and Gosselink, J.G. 1989. Faunal changes and bottomland hardwood forest loss in the Tensas watershed, Louisiana. *Conservation Biology* **3**: 282–292.

Burns, L.A. 1978. Productivity, Biomass, and Water Relations in a Florida Cypress Forest. Ph.D. dissertation, University of North Carolina, Chapel Hill.

Burns, P.Y. 1940. Ecological studies in an eastern Oklahoma flood plain. *Proceedings of the Oklahoma Academy of Science* **20**: 49–52.

Burrows, F.M. 1986. Aerial motion. In D.R. Murray (ed.) *Seed Dispersal*, pp. 1–47. Academic Press, Sydney.

Busch, D.E. 1984. Environmental data system of the Bureau of Reclamation. In R.E. Warner and K.M. Hendrix (eds.) *California Riparian Systems*, pp. 335–9. University of California Press, Berkeley.

Bush, J.K. and Van Auken, O.W. 1984. Woody-species composition of the upper San Antonio River gallery forest. *Texas Journal of Science* **36**: 139–48.

Butler, D.R. 1979. Dendrogeomorphological analysis of flooding and mass movement, Ram Plateau, Mackenzie Mountains, Northwest Territories. *Canadian Geographer* **23**: 62–5.

Butler, D.R. 1989. The failure of beaver dams and resulting outburst flooding: a geomorphic hazard of the southeastern Piedmont. *Geographical Bulletin* **31**: 29–38.

Butler, D.R. 1991*a*. The reintroduction of the beaver into the South. *Southeastern Geographer* **31**: 39–43.

Butler, D.R. 1991*b*. Beavers as agents of biogeomorphic change: a review and suggestions for teaching exercises. *Journal of Geography* **90**: 210–17.

Butler, D.R. and Malanson, G.P. 1990. Non-equilibrium geomorphic processes and patterns on avalanche paths in the northern Rocky Mountains, U.S.A. *Zeitschrift für Geomorphologie* **34**: 257–70.

Butler, D.R. and Malanson, G.P. 1993. Characteristics of two landslide-dammed lakes in a glaciated alpine environment. *Limnology and Oceanography*, in press.

Butler, D.R. and Walsh, S.J. 1990. Lithologic, structural, and topographic influences on snow-avalanche path location, eastern Glacier National Park, Montana. *Annals of the Association of American Geographers* **80**: 362–78.

Butler, D.R., Malanson, G.P. and Oelfke, J.G. 1987. Tree-ring analysis and natural hazard chronologies: minimum sample size and index values. *Professional Geographer* **39**: 41–7.

Butler, D.R., Malanson, G.P. and Walsh, S.J. 1991. Identification of deltaic wetlands at montane finger lakes. *Environmental Professional* **13**: 352–62.

Butler, D.R., Malanson, G.P. and Kupfer, J.A. 1992. Beaver, treefall, and cutbank erosion in midwestern rivers. *Abstracts, Annual Meeting of the Association of American Geographers*, pp. 29–30.

Cairns, J., Stauffer, J.R. and Hocutt, C.H. 1979. Opportunities for maintenance and rehabilitation of riparian habitats: eastern United States. In R.R. Johnson and J.F. McCormick (eds.) *Strategies for Protection and Management of Floodplain Wetlands and Other Riparian Ecosystems*, pp. 304–17. USA Forest Service General Technical Report WO-12.

Campbell, C.J. 1970. Ecological implications of riparian vegetation management. *Journal of Soil and Water Conservation* **25**: 49–52.

Campbell, C.J. and Dick-Peddie, W.A. 1964. Comparison of phreatophyte communities on the Rio Grande in New Mexico. *Ecology* **45**: 492–502.

Campbell, C.J. and Green, W. 1968: Perpetual succession of stream-channel vegetation in a semi-arid region. *Arizona Academy of Science* **5**: 86–98.

Campbell, D.G., Stone, J.L. and Rosas, A. 1992. A comparison of the phytosociology and dynamics of three floodplain (Varzea) forests of known ages, Rio Jurua, western Brazilian Amazon. *Botanical Journal of the Linnean Society* **108**: 213–37.

Campbell, K, and Frailey, C.D. 1984. Holocene flooding and species diversity in southwestern Amazonia. *Quaternary Research* **21**: 369–75.

Carbiener, R. and Schnitzler, A. 1990. Evolution of major pattern models and processes of alluvial forest of the Rhine in the rift valley (France/Germany). *Vegetatio* **88**: 115–29.

Carbiener, R. and Tremolieres, M. 1990. The Rhine rift valley groundwater – river interactions: evolution of their susceptibility to pollution. *Regulated Rivers: Research and Management* **5**: 375–89.

Cargill, S.M. 1988. Establishment of Native Plants on Disturbed Sites in Arctic Alaska. Ph.D. dissertation, University of Alaska, Fairbanks.

Carlson, J.Y., Andrus, C.W. and Froehlich, H.A. 1990. Woody debris, channel features, and macroinvertebrates of streams with logged and undisturbed riparian timber in northeastern Oregon, USA. *Canadian Journal of Fisheries and Aquatic Sciences* **47**: 1103–11.

Carrel, G. and Juget, J. 1987. La Morte du Sauget, un ancien méandre du Rhône: bilan hydrologique et biogéochimique. *Schweizerische Zeitschrift für Hydrologie* **49**: 102–25.

Castella, C. and Amoros, C. 1984. Repartition des characées dans les bras morts du Haut-Rhône et de l'Ain et signification écologique. *Cryptogamie, Algologie* **5**: 127–39.

Castella, C. and Amoros, C. 1986. Diagnostic phyto-écologique sur les anciens méandres. *Documents de Cartographie Ecologique* **29**: 97–108.

Castella, C. and Amoros, C. 1988. Freshwater macroinvertebrates as functional describers of the dynamics of former river beds. *Verhandlungen Internationale Vereiningung fur Theoretische und Angewandte Limnologie* **23**: 1299–305.

Castella, E., Richoux, P., Richardot-Coulet, M. and Roux, C. 1986. Un diagnostic écologique de trois anciens méandres base sur l'utilisation de descripteurs faunistiques. *Documents de Cartographie Ecologique* **29**: 109–122.

Caswell, H. 1988. Theory and models in ecology: a different perspective. *Ecological Modelling* **43**: 33–44.

Chambless, L.F. and Nixon, E.S. 1975. Woody vegetation-soil relations in a bottomland forest of east Texas. *Texas Journal of Science* **26**: 407–16.

Chapman, R.J., Hinckley, T.M., Lee, L.C. and Teskey, R.O. 1982. Impact of water level changes on woody riparian and wetland communities. Vol. X. Index and addendum to Vols. I-VIII. USA Fish and Wildlife Service OBS-82/23.

Chauvet, E. 1988. Influence of the environment on willow leaf litter decomposition in the alluvial corridor of the Garonne River. *Archiv für Hydrobiologie* **112**: 371–86.

Chauvet, E. and Decamps, H. 1989. Lateral interactions in a fluvial landscape: the River Garonne, France. *Journal of the North American Benthological Society* **8**: 9–17.

Chauvet, E. and Jean-Louis, A.M. 1988. Production de litiére de la ripisylve de la Garonne et apport au fleuve. *Oecologica Generalis* **9**: 265–79.

Chauvet, E. and Merce, J. 1988. Hypomycetes aquatiques: importance dans la decomposition des litiéres. *Revue des Sciences de l'Eau* **1**: 203–216.

Chavan, A.R. and Sabnis, S.D. 1960. Along the banks of the River Vishwamitri. *Indian Forestry* **86**: 469–74.

Childers, D.L. and Gosselink, J.G. 1990. Assessment of cumulative impacts to water quality in a forested wetland landscape. *Journal of Environmental Quality* **19**: 455–64.

Chorley, R.J. and Kennedy, B.A. 1971. *Physical Geography, A Systems Approach.* Prentice Hall International, London.

Chrisman, N.R. 1984. The role of quality information in the longterm functioning of a geographic information system. *Cartographica* **21**: 79–87.

Christensen, E.M. 1962. The rate of naturalization of tamarisk in Utah. *American Midland Naturalist* **68**: 51–7.

Christensen, E.M. 1963. Naturalization of Russian-olive (*Eleagnus angustifolia*) in Utah. *American Midland Naturalist* **70**: 133–7.

Chumbley, C.A., Baker, R.G. and Bettis, E.A. 1990. Midwestern Holocene paleoenvironments revealed by floodplain deposits in northeastern Iowa. *Science* **249**: 272–4.

Church, M. and Ryder, J.M. 1972. Paraglacial sedimentation: a consideration of fluvial processes conditioned by glaciation. *Geological Society of America Bulletin* **83**: 3059–3072.

Church, M. and Slaymaker, O. 1989. Disequilibrium of Holocene sediment yield in glaciated British Columbia. *Nature* **337**: 452–4.

Church, M., Kellerhals, R. and Day, T.J. 1989. Regional clastic sediment yield in British Columbia. *Canadian Journal of Earth Sciences* **26**: 31–45.

Clark, E.H., Haverkamp, J.A. and Chapman, W. 1985. *Eroding Soils: The Off-Farm Impacts*. The Conservation Foundation, Washington.

Clark, J.R. 1980. River corridor approach to bottomland management. *Transactions of the North American Wildlife and Natural Resources Conference* **45**: 387–91.

Clewell, A.F. and Lea, R. 1990. Creation and restoration of forested wetland vegetation in the southeastern United States. In *Wetland Creation and Restoration: The Status of the Science*, pp. 195–231. Island Press, Covelo, CA.

Clover, E.U. and Jotter, L. 1944. Floristic studies in the canyon of the Colorado and its tributaries. *American Midland Naturalist* **32**: 591–642.

Coffin, D.P. and Lauenroth, W.K. 1989. Disturbances and gap dynamics in a semiarid grassland: a landscape-level approach. *Landscape Ecology* **3**: 19–27.

Cohen, D. and Levin, S.A. 1991. Dispersal in patchy environments: the effects of temporal and spatial structure. *Theoretical Population Biology* **39**: 63–99.

Cohen, W.L., Hug, A.W., Taddese, A. and Cook, K.A. 1991. FACTA 1990 conservation and environmental highlights. *Journal of Soil and Water Conservation* **46**: 20–2.

Cole, D.N. and Marion, J.L. 1988. Recreation impacts in some riparian forests of the eastern United States. *Environmental Management* **12**: 99–107.

Collins, S.L., Risser, P.G. and Rice, E.L. 1981. Ordination and classification of mature bottomland forests in North Central Oklahoma. *Bulletin of the Torrey Botanical Club* **108**: 152–65.

Colyn, M., Gautier-Hion, A. and Verheyen, W. 1991. A re-appraisal of palaeoenvironmental history in Central Africa: evidence for a major fluvial refuge in the Zaire Basin. *Journal of Biogeography* **18**: 403–7.

Conchou, O. and Pautou, G. 1987. Modes of colonization of an heterogeneous alluvial area on the edge of the Garonne River by *Phalaris arundinacea* L. *Regulated Rivers* **1**. 37–48.

Connell, J.H. 1978. Diversity in tropical rain forests and coral reefs. *Science* **199**: 1302–10.

Connell, J.H. 1980. Diversity and the coevolution of competitors, or the ghost of competition past. *Oikos* **35**: 131–8.

Conner, W.H. and Day, J.W. Jr. 1976. Productivity and composition of a baldcypress-water tupelo site and a bottomland hardwood site in a Louisiana swamp. *American Journal of Botany* **63**: 1354–64.

Conner, W.H., Gosselink, J.G. and Parrando, R.T. 1981. Comparison of the vegetation of three Louisiana swamp sites with different flooding regimes. *American Journal of Botany* **63**: 320–31.

Cooper, C.M. and McHenry, J.R. 1989. Sediment accumulation and its effects on a Mississippi River oxbow lake. *Environmental Geology and Water Science* **13**: 33–7.

Cooper, J.R. and Gilliam, J.W. 1987. Phosphorous redistribution from cultivated fields into riparian areas. *Soil Science Society of America Journal* **51**: 1600–4.

Cooper, J.R., Gilliam, J.W. and Jacobs, T.C. 1986. Riparian areas as a control of nonpoint pollutants. In D.L. Correll (ed.) *Watershed Research Perspectives*, pp. 166–92. Smithsonian Institution, Washington.

Cooper, J.R., Gilliam, J.W., Daniels, R.B. and Robarge, W.P. 1987. Riparian areas as filters for agricultural sediment. *Soil Science Society of America Journal* **51**: 416–20.

Correll, D.L., Jordan, T.E. and Weller, D.E. 1991. Nutrient dynamics in coastal plain agricultural watersheds: the role of riparian forests. In *World Congress of Landscape Ecology*, p. 40. Carleton University, Ottawa.

Council on Environmental Quality. 1973. *Report on Channel Modifications*, Vol. I. Washington, Executive Office of the President.

Courtois, L.A. 1984. Temporal desert riparian systems – the Mojave River as an example. In R.E. Warner and K.M. Hendrix (eds.) *California Riparian Systems*, pp. 688–93. University of California Press, Berkeley.

Craig, M.R. 1992. Colonization on Point Bars by Woody Riparian Species. MA thesis, University of Iowa, Iowa City.

Crampton, C.B. 1987: Soils, vegetation and permafrost across an active meander of Indian River, central Yukon, Canada. *Catena* **14**: 157–63.

Crance, J.H. and Ischinger, L.S. 1989. Fishery functions and values of forested riparian wetlands. In *Water: Laws and Management*, pp. 14A9–16. American Water Resources Association, Bethesda, Maryland.

Crawford, N.H. and Linsley, R.K. 1966. Digital simulation in hydrology: Stanford Watershed Model IV. Stanford University Department of Civil Engineering Technical Report 39.

Crites, R.W. and Ebinger, J.E. 1969. Vegetation survey of floodplain forests in east-central Illinois. *Transactions of the Illinois Academy of Science* **62**: 316–30.

Croonquist, M.J. and Brooks, R.P. 1991. Use of avian and mammalian guilds as indicators of cumulative impacts in riparian-wetland areas. *Environmental Management* **15**: 701–14.

Crouch, G.L. 1979*a*. Changes in the vegetation complex of a cottonwood ecosystem on the South Platte River. *Great Plains Agricultural Council* **91**: 19–22.

Crouch, G.L. 1979*b*. Long-term changes in cottonwoods on a grazed and an ungrazed plains bottomland in northeastern Colorado. USA Forest Service Research Note RM-370.

Cummins, K.W., Wilzbach, M.A., Gates, D.M., Perry, J.B. and Taliaferro, W.B. 1989. Shredders and riparian vegetation. *BioScience* **39**: 24–30.

Currier, P.J. 1982. The Floodplain Vegetation of the Platte River: Phytosociology, Forest Development, and Seedling Establishment. Ph.D. dissertation, Iowa State University, Ames.

Curry, P. and Slater, F.M. 1986. A classification of river corridor vegetation from four catchments in Wales. *Journal of Biogeography* **13**: 119–32.

Dahm, C.N. and Sedell, J.R. 1986. The role of beaver on nutrient cycling in streams. *Journal of the Colorado-Wyoming Academy of Science* **18**: 32.

Dahlskog, S. 1966: Sedimentation and vegetation in a Lapland mountain delta. *Geografiska Annaler* **48**A, 86–101.

Dains, V.I. 1989. Water relations of white alder. In D.L. Abell (ed.) *Proceedings of the*

California Riparian Systems Conference. US Forest Service General Technical Report PSW-110, pp. 375–80.

Daniels, R.B. 1960. Entrenchment of the Willow Drainage Ditch, Harrison county, Iowa. *American Journal of Science* **258**: 167–76.

Dansereau, P. 1964. Six problems in New Zealand vegetation. *Bulletin of the Torrey Botanical Club* **91**: 114–40.

Dansereau, P. and Lems, K. 1957. *The Grading of Dispersal Types in Plant Communities and Their Ecological Significance*. Institut Botanique de l'Université de Montreal, Montreal.

Darley-Hill, S. and Johnson, W.C. 1981. Acorn dispersal by the blue jay (*Cyanocitta cristata*). *Oecologia* **50**: 231–2.

Darwin, C. 1859. *On the Origin of Species*. John Murray, London.

Davenport, D.C., Martin, P.E. and Hagan, R.M. 1982. Evapotranspiration from riparian vegetation: water relations and irrecoverable losses for saltcedar. *Journal of Soil and Water Conservation* **37**: 233–6.

Davis, M.B. 1981. Quaternary history and the stability of forest communities. In D.C. West, H.H. Shugart and D.B. Botkin (eds.) *Forest Succession*, pp. 132–53. Springer-Verlag, New York.

Dawdy, D.R. 1988. Natural consumptive use by tules, grasslands, and riparian forests, California Central Valley. In *Water-Use Data for Water Resources Management*, pp. 621–31. American Water Resources Association, Bethesda, Maryland.

Dawson, F.H. 1976. Organic contribution of stream edge forest litter fall to the chalk stream ecosystem. *Oikos* **27**: 13–18.

Dawson, T.E. and Ehleringer, J.R. 1991. Streamside trees that do not use stream water. *Nature* **350**: 335–7.

Day, F.P. 1987. Effects of flooding and nutrient enrichment on biomass allocation in *Acer rubrum* seedlings. *American Journal of Botany* **74**: 1541–54.

Day, R.T., Keddy, P.A., McNeill, J. and Carleton, T. 1988. Fertility and disturbance gradients: a summary model for riverine marsh vegetation. *Ecology* **69**: 1044–54.

DeAngelis, D.L. and Waterhouse, J.C. 1987. Equilibrium and nonequilibrium concepts in ecological models. *Ecological Monographs* **57**: 1–21.

DeByle, N.V. 1985. Wildlife. In N.V. DeByle and R.P. Winokur (eds.) *Aspen: Ecology and Management in the Western United States*. USA Forest Service General Technical Report RM-119, pp. 135–152.

Decamps, H. 1984. Towards a landscape ecology of river valleys. In J.H. Cooley and F.B. Golley (eds.) *Trends in Ecological Research for the 1980's*, pp. 163–78. New York: Plenum Press.

Decamps, H. and Fortune, M. 1991. Long-term ecological research and fluvial landscapes. In P.G. Risser (ed.) *Long-Term Ecological Research*, pp. 135–51. J. Wiley, New York.

Decamps, H. and Naiman, R.J. 1989. L'écologie des fleuves. *La Réchérche* **20**: 310–19.

Decamps, H., Fortune, M., Gazelle, F. and Pautou, G. 1988. Historical influence of man on the riparian dynamics of a fluvial landscape. *Landscape Ecology* **1**: 163–73.

Decamps, H., Joachim, J. and Lauga, J. 1987. The importance for birds of the riparian woodlands within the alluvial corridor of the River Garonne, S.W. France. *Regulated Rivers: Research and Management* **1**: 301–16.

Decamps, H., Muller, E., Blasco, F. and Lauga, J. 1991. Utilization of remote sensing in the ecology of large river valleys. *Verhandlungen der Internationale Vereiningung für Theoretische und Angewandte Limnologie* **24**: 2031–4.

Demaree, D. 1932. Submerging experiments with *Taxodium*. *Ecology* **13**: 258–62.

Demissie, M. and Bhowmik, N.G. 1987. Long-term impacts of river basin development on lake sedimentation: the case of Peoria Lake. *Water International* **12**: 23–32.

Denayer-De Smet, S. 1970. Biomasse, productivité et phytogeochimie de la végétation riveraine d'un ruisseau Ardennais. II. Aperçu phytogéochimique. *Bulletin de la Société Royale Botanique Belgique* **103**: 383–96.

Desaigues, B. 1990. The socio-economic value of ecotones, In R.J. Naiman and H. Decamps (eds.) *The Ecology and Management of Aquatic-Terrestrial Ecotones*, pp. 263–93. Man and the Biosphere Series Vol. 4. UNESCO, Paris.

Diamond, J.M. 1975. Assembly of species communities. In M.L. Cody and J.M. Diamond (eds.) *Ecology and Evolution of Communities*, Belknap Press, Cambridge, Massachusetts.

Dickson, R.E., Hosner, J.F. and Hosley, N.W. 1965. The effects of four water regimes upon the growth of four bottomland tree species. *Forest Science* **11**: 299–305.

Dietz, R.A. 1952. The evolution of a gravel bar. *Annals of the Missouri Botanical Garden* **39**: 249–54.

Dillaha, T.A., Reneau, R.B., Mostaghimi, S. and Lee, D. 1989. Vegetative filter strips for agricultural nonpoint source pollution control. *Transactions of the American Society of Agricultural Engineers (ASAE)* **32**: 513–9.

Dinerstein, E. 1991. Effects of *Rhinoceros unicornis* on riverine forest structure in lowland Nepal. *Ecology* **73**: 701–4.

Dirschl, H.J. and Coupland, R.T. 1972. Vegetation patterns and site relationships on the Saskatchewan River delta. *Canadian Journal of Botany* **50**: 647–75.

Dobrowski, K.A. 1964. Studies on the ecology of birds of the Vistula River. *Ekologia Polska* **12**: 615–51.

Dobzhansky, T. and Wright, S. 1943. Genetics of natural populations. X. Dispersion rates in *Drosophila pseudoobscura*. *Genetics* **28**: 304–40.

Dole, M.-J. 1983. Le domaine aquatiqué souterrain de la plaine alluviale du Rhône á l'est de Lyon. 1. Diversité hydrologique et biocenotique de trois stations répésentatives de la dynamique fluviale. *Vie Milieu* **33**: 219–29.

Dole, M.-J. 1984. Structure biocénotique des niveaux superieurs de la nappe alluviale du Rhône á l'est de Lyon. *Memoires Biospeliologie* **9**: 17–26.

Donovan, L.A. and McLeod, K.W. 1984. Bald cypress seedling response to thermal and hydrological regimes. *Bulletin of the Ecological Society of America* **65**: 209.

Donovan, L.A., McLeod, K.W., Sherrod, K.C. Jr. and Stumpff, N.J. 1988. Response of woody swamp seedlings to flooding and increased water temperatures. I. Growth, biomass, and survivorship. *American Journal of Botany* **75**: 1181–90.

Doty, C.W., Parsons, J.E., Nassehzadeh-Tabrizi, A., Skaggs, R.W. and Badr, A.W. 1984. Stream water levels affect field water tables and corn yields. *Transactions of the American Society of Agricultural Engineers (ASAE)* **27**: 1300–6.

Douglas, D.A. 1987. Growth of *Salix setchelliana* on a Kluane River point bar, Yukon Territory, Canada. *Arctic and Alpine Research* **19**: 35–44.

Douglas, D.A. 1989. Clonal growth of *Salix setchelliana* on glacial river gravel bars in Alaska. *Journal of Ecology* **77**: 112–26.

Douthwaite, R.J. 1987. Lowland forest resources and their conservation in southern Somalia. *Environmental Conservation* **14**: 29–35.

Drury, W.H. 1956. Bog flats and physiographic processes in the upper Kuskokwim River region, Alaska. *Contributions of the Gray Herbarium* **178**: 1–130.

DuBarry, A.P. 1963. Germination of bottomland tree seed while immersed in water. *Journal of Forestry* **61**: 225–6.

Duever, M.J., Carlson, J.E. and Riopelle, L.A. 1975. Ecosystem analysis at Corkscrew Swamp. In H.T. Odum and K.C. Ewel (eds.) *Cypress Wetlands for Water Management, Recycling and Conservation*, pp. 627–725. University of Florida, Gainesville.

Duever, M.J., Carlson, J.E. and Riopelle, L.A. 1984. Corkscrew Swamp: a virgin cypress stand. In K.C. Ewel and H.T. Odum (eds.) *Cypress Swamps*, pp. 334–48. University Presses of Florida, Gainesville.

Duijsings, J.J.H.M. 1987. A sediment budget for a forested catchment in Luxembourg and its implications for channel development. *Earth Surface Processes and Landforms* **12**: 173–84.

Dumont, J.F., Lamotte, S. and Kahn, F. 1990. Wetland and upland forest ecosystems in Peruvian Amazonia: plant species diversity in the light of some geological and botanical evidence. *Forest Ecology and Management* **33/34**: 125–39.

Duncan, S.L. 1984. Leaving it to beaver. *Environment* **26**(3): 41–5.

Dunham, K.M. 1989*a*. Vegetation-environment relations of a Middle Zambezi floodplain. *Vegetatio* **82**: 13–24.

Dunham, K.M. 1989*b*. Long-term changes in Zambezi riparian woodlands, as revealed by photopanoramas. *African Journal of Ecology* **27**: 263–75.

Dunham, K.M. 1990*a*. Biomass dynamics of herbaceous vegetation in Zambezi riverine woodlands. *African Journal of Ecology* **28**: 200–12.

Dunham, K.M. 1990*b*. Fruit production by *Acacia albida* trees in Zambezi riverine woodlands. *Journal of Tropical Ecology* **6**: 445–457.

Du Toit, R.F. 1984. Some environmental aspects of proposed hydroelectric schemes on the Zambezi River, Zimbabwe. *Biological Conservation* **28**: 73–87.

Duvigneaud, P. and Denayer-De Smet, S. 1970. Biomasse, productivité et phytogéochimie de la végétation riveraine d'un ruisseau Ardennais. I. Aperçu sur les sols, la végétation et la biomasse de la strate au sol. *Bulletin de la Société Royale Botanique Belgique* **103**: 353–82.

Eckblad, J.W., Peterson, N.L. and Ostlie, K. 1977. The morphometry, benthos and sedimentation rates of a floodplain lake in Pool 9 of the upper Mississippi River. *American Midland Naturalist* **97**: 433–43.

Eckersten, H. 1986. Simulated willow growth and transpiration: the effect of high and low resolution weather data. *Agricultural and Forest Meteorology* **38**: 289–306.

Ellenberg, H. 1987. Vegetation und Mikroklima in einem Waldbachtal der Alpen. *Botanische Jahrbücher für Systematik* **108**: 499–514.

Ellery, W.N., Ellery, K., Rogers, K.H., McCarthy, T.S. and Walker, B.H. 1990. Vegetation of channels of the northeastern Okavango Delta, Botswana. *African Journal of Ecology* **28**: 276–90.

Ellis, R.W. and Whelan, J.B. 1979. Impact of stream channelization on riparian small mammal and bird populations in Piedmont Virginia. *Transactions of the Northeast Section, The Wildlife Society* **35**: 92–104.

Elwood, J.W., Newbold, J.D., O'Neill, R.V. and Van Winkle, W. 1983. Resource spiraling: an operational paradigm for analyzing lotic ecosystems. In T.D. Fontaine and S.M. Bartell (eds.) *Dynamics of Lotic Ecosystems*, pp. 3–27. Ann Arbor Science, Ann Arbor.

Endler, J.A. 1977. *Geographic Variation, Speciation, and Clines*. Princeton University Press.

Ericsson, K.A. and Schimpf, D.J. 1986. Woody riparian vegetation of a Lake Superior tributary. *Canadian Journal of Botany* **64**: 769–73.

Evenden, A.G. 1989. Ecology and Distribution of Riparian Vegetation in the Trout Creek Mountains of Southeastern Oregon. Ph.D. dissertation, Oregon State University, Corvallis.

Everitt, B.L. 1980. Ecology of saltcedar – a plea for research. *Environmental Geology* **3**: 77–84.

Fagerstrom, T. and Agren, G.I. 1979. Theory for coexistence of species differing in regeneration properties. *Oikos* **33**: 1–10.

Fagot, P., Gadiolet, P., Magne, M. and Bravard, J.P. 1989. Une étude dendrochronologique dans le lit majeur de l'Ain: la forêt alluviale comme descripteur d'une 'metamorphose fluviale'. *Revue de Géographie de Lyon* **64**: 213–23.

Fail, J.L., Haines, B.L. and Todd, R.L. 1987. Riparian forest communities and their role in nutrient conservation in an agricultural watershed. *American Journal of Alternative Agriculture* **2**: 114–21.

Fanshawe, D.B. 1954. Riparian vegetation of British Guiana. *Journal of Ecology* **42**: 289–95.

Farjon, A. and Bogaers, P. 1985. Vegetation and primary succession along the Porcupine River in interior Alaska. *Phytocoenologia* **13**: 465–504.

Farrell, J.A.K. 1968. Preliminary notes on the vegetation of the Lower Sabi-Lundi Basin, Rhodesia. *Kirkia* **6**: 223–248.

Featherly, H.I. 1940. Silting and forest succession on Deep Fork in southwestern Creek Count, Oklahoma. *Proceedings of the Oklahoma Academy of Science* **20**: 63–4.

Fenner, P., Brady, W.W. and Patton, D.R. 1985. Effects of regulated water flows on regeneration of Fremont cottonwood. *Journal of Range Management* **38**: 135–8.

Fiebig, D.M., Lock, M.A. and Neal, C. 1990. Soil water in the riparian zone as a source of carbon for a headwater stream. *Journal of Hydrology* **116**: 217–37.

Finch, D.M. 1991. Positive associations among riparian bird species correspond to elevational changes in plant communities. *Canadian Journal of Zoology* **69**: 951–63.

Finlayson, C.M. 1991. Production and major nutrient composition of three grass species on the Magela floodplain, Northern Territory, Australia. *Aquatic Botany* **41**: 263–80.

Finlayson, C.M., Crowie, I.D. and Bailey, B.J. 1990. Sediment seedbanks in grasslands on the Magela Creek floodplain, northern Australia. *Aquatic Botany* **38**: 163–76.

Firth, P.L. and Hooker, K.L. 1989. Plant community structure in disturbed and undisturbed forested wetlands. In D.W. Fisk (ed.) *Wetlands: Concerns and Successes*, pp. 101–13. American Water Resources Association, Bethesda, Maryland.

Fitzpatrick, T.J. and Fitzpatrick, M.F.L. 1902. A study of the island flora of the Mississippi River near Sabula, Iowa. *Plant World* **5**: 198–201.

Fleshman, C. and Kaufman, D.S. 1984. The South Fork (Kern River) wildlife area: will the commitment be forgotten? In R.E. Warner and K.M. Hendrix (eds.) *California Riparian Systems*, pp. 58–67. University of California Press, Berkeley.

Fonda, R.W. 1974. Forest succession in relation to river terrace development in Olympic National Park, Washington. *Ecology* **55**: 927–42.

Ford, A.L. and Van Auken, O.W. 1982. The distribution of woody species in the Guadalupe River floodplain forest in the Edwards Plateau of Texas. *Southwestern Naturalist* **27**: 383–92.

Forman, R.T.T. 1979. *Pine Barrens: Ecology and Landscape*. Academic Press, New York.

Forman, R.T.T. 1983. Corridors in a landscape: their ecological structure and function. *Ekologia (CSSR)* **2**: 375–87.

Forman, R.T.T. 1990. The beginnings of landscape ecology in America. In I.S. Zonneveld and R.T.T. Forman (eds.) *Changing Landscapes: An Ecological Perspective*, pp. 35–41. Springer-Verlag, New York.

Forman, R.T.T. and Godron, M. 1981. Patches and structural components for a landscape ecology. *BioScience* **31**: 733–40.

Forman, R.T.T. and Godron, M. 1986. *Landscape Ecology*. Wiley, New York.

Forsberg, B.R., Devol, A.H., Richey, J.E., Martinelli, L.A. and dos Santos, H. 1988. Factors controlling nutrient concentrations in Amazon floodplain lakes. *Limnology and Oceanography* **33**: 41–56.

Fortune, M. 1988. Historical changes of a large river in an urban area: the Garonne River, Toulouse, France. *Regulated Rivers: Research and Management* **2**: 179–86.

Francis, M.M., Naiman, R.J. and Melillo, J.M. 1985. Nitrogen fixation in subarctic streams influenced by beaver (*Castor canadensis*). *Hydrobiologia* **121**: 193–202.

Frangi, J.L. and Lugo, A.E. 1985. Ecosystem dynamics of a subtropical floodplain forest. *Ecological Monographs* **55**: 351–69.

Franz, E.H. and Bazzaz, F.A. 1977. Simulation of vegetation response to modified hydrologic regimes: a probabilistic model based on niche differentiation in a floodplain forest. *Ecology* **58**: 176–83.

Fredrickson, L.H. 1979. Floral and faunal changes in lowland hardwood forests in Missouri resulting from channelization, drainage, and impoundment. US Fish and Wildlife Service OBS-78/91.

Freeman, C.E. and Dick-Peddie, W.A. 1970. Woody riparian vegetation in the Black and Sacramento mountain Ranges, southern New Mexico. *Southwestern Naturalist* **15**: 145–64.

Frenette, M. and Julien, P.Y. 1986. LAVSED-I – Un modèle pour prédire l'erosion des bassins et le transfert de sédiment fins dans les cours d'eau nordiques. *Canadian Journal of Civil Engineering* **13**: 150–61.

Frost, I. and Miller, M.C. 1987. Late Holocene flooding in the Ecuadorian rain forest. *Freshwater Biology* **18**: 443–53.

Frye, R.J. and Quinn, J.A. 1979. Forest development in relation to topography and soils in a flood-plain of the Raritan River, New Jersey. *Bulletin of the Torrey Botanical Club* **106**: 334–45.

Furch, K., Junk, W.J. and Campos, Z.E.S. 1989. Nutrient dynamics of decomposing leaves from Amazonian floodplain forest species in water. *Amazonia* **11**: 91–116.

Furness, H.D. and Breen, C.M. 1980. The vegetation of seasonally flooded areas of the Pongolo River floodplain. *Bothalia* **13**: 217–31.

Fustec, E., Mariotti, A., Grillo, X. and Sajus, J. 1991. Nitrate removal by denitrification in alluvial ground water: role of a former channel. *Journal of Hydrology* **123**: 337–54.

Fyles, J.W. and Bell, M.A.M. 1986. Vegetation colonizing river gravel bars in the Rocky Mountains of southeastern British Columbia. *Northwest Science* **60**: 8–14.

Galiano, E.F., Sterling, A. and Viejo, J.L. 1985. The role of riparian forests in the conservation of butterflies in a Mediterranean area. *Environmental Conservation* **12**: 361–2.

Ganeshaiah, K.N. and Shaanker, R.U. 1991. Seed size optimization in a wind dispersed tree *Butea monosperma*: a trade-off between seedling establishment and pod dispersal efficiency. *Oikos* **60**: 3–6.

Gareera, L.M. and Naumova L.G. 1980. Experiment on quantitative substantiation of territorial units of river floodplain vegetation in the Mongolian People's Republic. *Soviet Journal of Ecology* **11**: 229–34.

Gastaldo, R.A., Bearce, S.C., Degges, C.W., Hunt, R.J., Peebles, M.W. and Violette, D.L. 1989. Biostratinomy of a Holocene oxbow lake: a backswamp to mid-channel transect. *Review of Palaeobotany and Palynology* **58**: 47–59.

Gatewood, J.S., Robinson, T.W., Colby, B.R., Hem, J.D. and Halpenny, L.C. 1950. Use of water by bottom-land vegetation in lower Safford Valley, Arizona. US Geological Survey Water Supply Paper 1103.

Gawler, S.C. 1988. Disturbance-Mediated Population Dynamics of *Pedicularis furbishiae* S. Wats., a Rare Riparian Endemic. Ph.D. dissertation, University of Wisconsin, Madison.

Gay, L.W. 1985. Evapotranspiration from saltcedar along the lower Colorado River. In R.R. Johnson, C.D. Ziebell, D.R. Patton, P.F. Ffolliott and R.H. Hamre (eds.) *Riparian Ecosystems and Their Management: Reconciling Conflicting Uses.* US Forest Service General Technical Report RM-120, pp. 171–4.

Gay, L.W. and Hartman, R.K. 1982. ET measurements over riparian saltcedar on the Colorado River. *Hydrology and Water Resources of Arizona and the Southwest* **12**: 9–15.

Gazelle, F. 1987. Recoupement de méandre á Sorde-l'Abbaye (Pyrénées-Atlantiques). *Revue Géographique des Pyrénées et du Sud-Ouest* **58**: 81–4.

Gazelle, F. 1989. Le rôle des Pyrénées dans l'abondance de la Garonne toulousaine. *Revue Géographique des Pyrénées et du Sud-Ouest* **60**: 503–20.

Gee, D.M., Anderson, M.G. and Barid, L. 1990. Large-scale floodplain modelling. *Earth Surface Processes and Landforms* **15**: 513–23.

Gehu, J.M. (ed.) 1980. *La Végétation des Forêts Alluviales. Colloques Phytosociologiques* **9**. J. Cramer, Vaduz.

Gehu, J.M. and Franck, J. 1980. Observations sur les saulaies riveraines de la vallée de la Loire, des sources á l'embouchure. *La Végétation des Forêts Alluviales. Colloques Phytosociologiques* **9**: 305–23.

Gemborys, S.R. and Hodgkins, E.J. 1971. Forests of small stream bottoms in the coastal plain of southwestern Alabama. *Ecology* **52**: 70–84.

Gesink, R.W., Tomanek, G.W. and Hulett, G.K. 1970. A descriptive survey of woody phreatophytes along the Arkansas River in Kansas. *Transactions of the Kansas Academy of Science* **73**: 55–9.

Giesel, J.T. 1976. Reproductive strategies as adaptations to life in temporally heterogeneous environments. *Annual Review of Ecology and Systematics* **7**: 57–79.

Gill, C.J. 1970. The flooding tolerance of woody species – a review. *Forestry Abstracts* **31**: 671–88.

Gill, D. 1972*a*. The point bar environment in the Mackenzie River delta. *Canadian Journal of Earth Science* **9**, 1382–93.

Gill, D. 1972*b*. The evolution of a discrete beaver habitat in the Mackenzie River delta, Northwest Territories. *Canadian Field-Naturalist* **86**: 233–9.

Gill, D. 1973: Floristics of a plant succession sequence in the Mackenzie delta, Northwest Territories. *Polarforschung* **43**, 55–65.

Gilliam, J.W., Skaggs, R.W. and Doty, C.W. 1986. Controlled agricultural drainage: an alternative to riparian vegetation. In D.L. Correll (ed.) *Watershed Research Perspectives*, pp. 225–43. Smithsonian Institution, Washington.

Girault, D. 1990. Piezometrical measurements as an aid in establishing a typology of forest communities growing on hydromorphic soils. *Vegetatio* **88**: 131–3.

Girel, J. 1986. Télédetection et cartographie à grand echelle de la vegetation alluviale: exemple de la basse plaine de l'Ain. *Documents de Cartographie Ecologique* **29**: 45–74.

Girel, J. and Manneville, O. 1991. Évolution de la végétation ripariale et palustre: les petits affluents rhodaniens du Jura méridional. *Bulletin Mensuel de la Société Linnéenne de Lyon* **60**: 112–27.

Girel, J. and Pautou, G. 1982. Les pelouses calcaires des alluvions de l'Ain en amont de la confluence avec le Rhône. *Les Pelouses Calcaires. Colloques Phytosociologiques* **11**: 229–41.

Girel, J., Pautou, G. and Pais, A. 1986. La végétation de la basse plaine de l'Ain. *Documents de Cartographie Ecologique* **29**: map supplement.

Glascock, S. and Ware, S. 1979. Forests of small stream bottoms in the peninsula of Virginia. *Virginia Journal of Science* **30**: 17–21.

Gleason, H.A. 1909. The vegetation history of a river dune. *Illinois Academy of Science Transactions* **2**: 19–26.

Gleason, H.A. 1922. The vegetational history of the Middle West. *Annals of the Association of American Geographers* **12**: 39–85.

Gleason, H.A. 1926. The individualistic concept of the plant association. *Bulletin of the Torrey Botanical Club* **53**: 7–26.

Glime, J.M. and Vitt, D.H. 1987. A comparison of bryophyte species diversity and niche structure of montane streams and stream banks. *Canadian Journal of Botany* **65**: 1824–37.

Good, B.J. and Whipple, S.A. 1982. Tree spatial patterns: South Carolina bottomland and swamp forests. *Bulletin of the Torrey Botanical Club* **109**: 529–36.

Goodall, D.W. 1974. Problems of scale and detail in ecological modelling. *Journal of Environmental Management* **2**: 149–57.

Gosselink, J.G., Shaffer, G.P., Lee, L.C., Burdick, D.M., Childers, D.L., Leibowitz, N.C., Hamilton, S.C., Boumans, R., Cushman, D., Fields, S., Kock, M. and Visser, J.M. 1990. Landscape conservation in a forested wetland watershed. *BioScience* **40**: 588–600.

Gottesberger, G. 1978. Seed dispersal by fish in the inundation regions of Humaita, Amazonia. *Biotropica* **10**: 170–83.

Gottesfeld, A.S. and Johnson Gottesfeld, L.M. 1990. Floodplain dynamics of a wandering river, dendrochronology of the Morice River, British Columbia, Canada. *Geomorphology* **3**: 159–79.

Goulding, M. 1980. *The Fisheries of the Forest.* University of California Press, Berkeley.

Grace, J.B. 1991. A clarification of the debate between Grime and Tilman. *Functional Ecology* **5**: 583–7.

Gradek, P., Saslaw, L. and Nelson, S. 1989. An application of BLM's riparian inventory procedure to rangeland riparian resources in the Kern and Kaweah River watersheds. In D.L. Abell (ed.) *Proceedings of the California Riparian Systems Conference.* USA Forest Service General Technical Report PSW-110, pp. 109–15.

Graf, W.L. 1978. Fluvial adjustments to the spread of tamarisk in the Colorado Plateau region. *Geological Society of America Bulletin* **89**: 1491–501.

Graf, W.L. 1980. Riparian management: a flood control perspective. *Journal of Soil and Water Conservation* **35**: 158–61.

Graf, W.L. 1982. Tamarisk and river-channel management. *Environmental Management* **6**: 283–96.

Graf, W.L. 1985. *The Colorado River.* Resource Publications in Geography. Association of American Geographers, Washington.

Graf, W.L., Trimble, S.W., Toy, T.J. and Costa, J.E. 1980. Geographic geomorphology in the eighties. *Professional Geographer* **32**: 279–84.

Grannemann, N.G. and Sharp, J.M. 1979. Alluvial hydrogeology of the lower Missouri River valley. *Journal of Hydrology* **40**: 85–99.

Green, J. 1972. Freshwater ecology in the Mato Grosso, central Brazil II. Associations of Cladocera in meander lakes of the Rio Suia Missu. *Journal of Natural History* **6**: 215–27.

Green, W.E. 1947. Effect of water impoundment on tree mortality and growth. *Journal of Forestry* **45**: 118–20.

Gregory, K.J. 1985. *The Nature of Physical Geography.* Edward Arnold, London.

Gregory, K.J., Gurnell, A.M. and Hill, C.T. 1985. The permanence of debris dams related to river channel processes. *Hydrological Sciences Journal* **30**: 371–81.

Gregory, S.V., Swanson, F.J., McKee, W.A. and Cummins, K.W. 1991. An ecosystem perspective of riparian zones. *BioScience* **41**: 540–51.

Griffin, G.F., Stafford Smith, D.M., Morton, S.R., Allan, G.E. and Masters, K.A. 1989. Status and implications of the invasion of Tamarisk (*Tamarix aphylla*) on the Finke River, Northern Territory, Australia. *Journal of Environmental Management* **29**: 297–315.

Grime, J.P. 1979. *Plant Strategies and Vegetation Processes.* Wiley, Chichester.

Groeneveld, D.P. and Griepentrog, T.E. 1985. Interdependence of groundwater, riparian vegetation, and streambank stability: a case study. In R.R. Johnson, C.D. Ziebell, D.R. Patton, P.F. Ffolliott and R.H. Hamre (eds.) *Riparian Ecosystems and Their Management: Reconciling Conflicting Uses.* USA Forest Service General Technical Report RM-120, pp. 44–8.

Grootjans, A.P., van Diggelen, R., Wassen, M.J. and Wiersinga, W.A. 1988. The effects of drainage on groundwater quality and plant species distribution in stream valley meadows. *Vegetatio* **75**: 37–48.

Grubaugh, J.W. and Anderson, R.V. 1988. Spatial and temporal availability of floodplain habitat: long-term changes at Pool 19, Mississippi River. *American Midland Naturalist* **119**: 402–11.

Grubaugh, J.W. and Anderson, R.V. 1989. Upper Mississippi River: seasonal and floodplain forest influences on organic matter transport. *Hydrobiologia* **174**: 235–44.

Grubb, P.J. 1977. The maintenance of species richness in plant communities: the importance of the regeneration niche. *Biological Reviews* **52**: 107–45.

Grunda, B. 1985. Activity of decomposers and processes of decomposition in soil. In M. Penka, M. Vyskot, E. Klimo and F. Vasicek (eds.) *Floodplain Forest Ecosystem. I. Before Water Management Measures*, pp. 389–414. Elsevier, Amsterdam.

Gruntfest, E. (ed.) 1991. *Multi-Objective River Corridor Planning*. Association of State Floodplain Managers, Madison, Wisconsin.

Gurnell, A.M. and Gregory, K.J. 1984. The influence of vegetation on stream channel processes. In T.P. Burt and D.E. Walling (eds.) *Catchment Experiments in Fluvial Geomorphology*, pp. 515–35. Norwich, Geo Books.

Gurtz, M.E. and Tate, C.M. 1988. Hydrologic influences on leaf decomposition in a channel and adjacent bank of a gallery forest ecosystem. *American Midland Naturalist* **120**: 11–21.

Gurtz, M.E., Marzolf, G.R., Killingbeck, K.T., Smith, D.L. and McArthur, J.V. 1988. Hydrologic and riparian influences on the import and storage of coarse particulate organic matter in a prairie stream. *Canadian Journal of Fisheries and Aquatic Sciences* **45**: 655–65.

Guy, P.R. 1981. River bank erosion in the mid-Zambezi valley, downstream of Lake Kariba. *Biological Conservation* **19**: 199–212.

Haase, G. 1964. Landschaftsökologische Detailuntersuchung und naturraumliche Gliederung. *Petermanns Geographische Mitteilungen* **109**: 8–30.

Hack, J.T. and Goodlett, J.C. 1960. Geomorphology and forest ecology of a mountain region in the central Appalachians. US Geological Survey Professional Paper 347.

Hadley, R.F. 1961. Influence of riparian vegetation on channel shape, northeastern Arizona. US Geological Survey Professional Paper 424-C, pp. 30–1.

Hall, T.F. and Penfound, W.T. 1939. A phytosociological study of cypress-gum swamp in southeastern Louisiana. *American Midland Naturalist* **21**: 378–95.

Hall, T.F. and Penfound, W.T. 1943. Cypress-gum communities in the Blue Girth Swamp near Selma, Alabama. *Ecology* **24**: 208–17.

Hallgren, S.W. 1989. Growth response of *Populus* hybrids to flooding. *Annales Sciences Forestieres* **46**: 361–72.

Hall-Martin, A.J. 1975. Classification and ordination of forest thicket vegetation of the Lengwe National Park, Malawi. *Kirkia* **10**: 131–84.

Halwagy, R. 1963. Studies on the succession of vegetation on some islands and sandbanks in the Nile near Khartoum, Sudan. *Vegetatio* **11**: 217–34.

Hamilton, K.E., Ferguson, A., Taggart, J.B., Tomasson, T., Walker, A. and Fahy, E. 1989. Post-glacial colonization of brown trout, *Salmo trutta* L., as a phylogeographic marker locus. *Journal of Fish Biology* **35**: 651–64.

Hammer, D.E. and Kadlec, R.H. 1986. A model for wetland surface water dynamics. *Water Resources Research* **22**: 1951–8.

Hanes, T.L., Friesen, R.D. and Keane, K. 1989. Alluvial scrub vegetation in coastal southern California. In D.L. Abell (ed.) *Proceedings of the California Riparian Systems Conference*. US Forest Service General Technical Report PSW-110, pp. 187–93.

Hansen, A.J., di Castri, F. and Naiman, R.J. 1988. Ecotones: what and why? *Biology International*, special issue **17**: 9–46.

Hanson, J.S., Malanson, G.P. and Armstrong, M.P. 1989. Spatial constraints on the response of forest vegetation to climate change. In G.P. Malanson (ed.) *Natural Areas Facing Climate Change*, pp. 1–23. SPB Academic, The Hague.

Hanson, J.S., Malanson, G.P. and Armstrong, M.P. 1990. Modelling the effects of dispersal and landscape fragmentation on forest dynamics. *Ecological Modelling* **49**: 277–96.

Hardin, E.D. and Wistendahl, W.A. 1983. The effects of floodplain trees on herbaceous vegetation patterns, microtopography, and litter. *Bulletin of the Torrey Botanical Club* **110**: 23–30.

Hardin, E.D., Lewis, K.P. and Wistendahl, W.A. 1989. Gradient analysis of floodplain forests along three rivers in unglaciated Ohio. *Bulletin of the Torrey Botanical Club* **116**: 258–64.

Hardiyanto, E.B. 1989. Genetic Variation in Two contrasting Habitats of Eastern Cottonwood Responses to Different Water Status and Nitrogen Levels. Ph.D. dissertation, Michigan State University, Lansing.

Harmon, M.E., Franklin, J.F., Swanson, F.J., Sollins, P., Gregory, S.V., Lattin, J.D., Anderson, N.H., Cline, S.P., Aumen, N.G., Sedell, J.R., Lienkaemper, G.W., Cromack, K. and Cummins, K.W. 1986. Ecology of coarse woody debris in temperate ecosystems. *Advances in Ecological Research* **15**: 133–302.

Harrington, C.A. 1987. Responses of red alder and black cottonwood seedlings to flooding. *Physiologia Plantarum* **69**: 35–48.

Harris, D.R. 1966. Recent plant invasions in the arid and semi-arid Southwest of the United States. *Annals of the Association of American Geographers* **56**: 408–22.

Harris, R.R. 1986. Occurrence patterns of riparian plants and their significance to water resource development. *Biological Conservation* **38**: 273–86.

Harris, R.R. 1987. Occurrence of vegetation on geomorphic surfaces in the active floodplain of a California alluvial stream. *American Midland Naturalist* **118**: 393–405.

Harris, R.R. 1988. Associations between stream valley geomorphology and riparian vegetation as a basis for landscape analysis in the eastern Sierra Nevada, California, USA. *Environmental Management* **12**: 219–28.

Harris, R.R., Fox, C.A. and Risser, R. 1987. Impacts of hydroelectric development on riparian vegetation in the Sierra Nevada region, California, USA. *Environmental Management* **11**: 519–27.

Hartland-Rowe, R. and Wright, P.B. 1975. Effects of sewage effluent on a swampland stream. *Verhandlungen Internationale Vereiningung für Theoretische und Angewandte Limnologie* **19**: 1575–83.

Hartshorne, R. 1939. *The Nature of Geography*. Association of American Geographers, Lancaster, Pennsylvania.

Harvey, D. 1969. *Explanation in Geography*. St. Martin's Press, New York.

Hawk, G.M. and Zobel, D.B. 1974. Forest succession on alluvial landforms of the

McKenzie River valley, Oregon. *Northwest Science* **48**, 245–265.
Hawkins, C.P., Murphy, M.L., Anderson, N.H. and Wilzbach, M.A. 1983. Density of fish and salamanders in relation to riparian canopy and physical habitat in streams of the northwestern United States. *Canadian Journal of Fisheries and Aquatic Sciences* **40**: 1173–85.
Heede, B.H. 1972. Flow and channel characteristics of two high mountain streams. US Forest Service Research Paper RM-96.
Heede, R. 1985. Interactions between streamside vegetation and stream dynamics. In R.R. Johnson, C.D. Ziebell, D.R. Patton, P.F. Ffolliott and R.H. Hamre (eds.) *Riparian Ecosystems and Their Management: Reconciling Conflicting Uses*. US Forest Service General Technical Report RM-120, pp. 54–8.
Hefley, H.M. 1937. Ecological studies on the Canadian River floodplain in Cleveland County, Oklahoma. *Ecological Monographs* **7**: 346–402.
Heger, L., Parker, M.L. and Kennedy, R.W. 1974. X-ray densitometry: a technique and an example of application. *Wood Science* **7**: 140–48.
Heimans, J. 1954. L'accessibilité, terme nouveau de phytogéographie. *Vegetatio* **5/6**: 142–6.
Hermy, M. 1980. A numerical approach to the phytosociology of riverine woods to the south of Bruges (Flanders, Belgium). *La Vegetation des Forets Alluviales. Colloques Phytosociologiques* **9**: 227–58.
Hermy, M. and Stieperaere, H. 1981. An indirect gradient analysis of the ecological relationships between ancient and recent riverine woodlands to the south of Bruges (Flanders, Belgium). *Vegetatio* **44**, 43–9.
Herrera, J. 1991. The reproductive biology of a riparian mediterranean shrub, *Nerium oleander* L. (Apocynaceae). *Botanical Journal of the Linnean Society* **106**: 147–72.
Hewitt, M.J. III. 1990. Synoptic inventory of riparian ecosystems: the utility of Landsat Thematic Mapper data. *Forest Ecology and Management* **33/34**: 605–20.
Hewlett, J.D. and Hibbert, A.R. 1967. Factors affecting the response of small watersheds. In W.E. Sopper and H.W. Lull (eds.) *Forest Hydrology*, pp. 275–90. Pergamon Press, London.
Hey, J. 1991. A multi-dimensional coalescent process applied to multi-allelic selection models and migration models. *Theoretical Population Biology* **39**: 30–48.
Hickin, E.J. 1974. The development of meanders in natural river-channels. *American Journal of Science* **274**: 414–42.
Hickin, E.J. 1984. Vegetation and river channel dynamics. *Canadian Geographer* **28**: 111–26.
Hicks, B.J., Beschta, R.L. and Harr, R.D. 1991. Long-term changes in streamflow following logging in western Oregon and associated fisheries implications. *Water Resources Bulletin* **27**: 217–26.
Hill, A.R. 1983. Denitrification – its importance in a river draining an intensively cropped watershed. *Agriculture, Ecosystems, and Environment* **10**: 47–62.
Hill, A.R. 1986. Stream nitrate-N loads in relation to variations in annual and seasonal runoff regimes. *Water Resources Bulletin* **22**: 829–39.
Hill, A.R. and Shacklet, M. 1989. Soil N-mineralization and nitrification in relation to nitrogen solution chemistry in a small forested watershed. *Biogeochemistry* **8**: 167–84.

Hill, A.R. and Warwick, J. 1987. Ammonium transformations in spring water within the riparian zone of a small woodland stream. *Canadian Journal of Fisheries and Aquatic Sciences* **44**: 1948–56.

Hill, B.H., Gardner, T.J. and Ekisola, O.F. 1988. Breakdown of gallery forest leaf litter in intermittent and perennial prairie streams. *Southwestern Naturalist* **33**: 323–31.

Holland, M.M. 1988. SCOPE/MAB technical consultations on landscape boundaries. Report of a SCOPE/MAB workshop on ecotones. *Biology International*, Special Issue **17**: 47–106.

Hollander, G. 1991. Rural fortunes: the battle over Brushy Creek. Unpublished research paper, Dept. of Geography, University of Iowa.

Holstein, G. 1984. Californian riparian forests: deciduous islands in an evergreen sea. In: R.E. Warner and K.M. Hendrix (eds.) *California Riparian Systems*, pp. 2–22. University of California Press, Berkeley.

Hook, D.D. 1984. Adaptations to flooding with fresh water. In T.T. Kozlowski (ed.) *Flooding and Plant Growth*, pp. 265–94. Academic Press, Orlando.

Hook, D.D. and Brown, C.L. 1973. Root adaptations and relative flood tolerance of five hardwood species. *Forest Science* **19**: 225–9.

Hook, D.D. and Scholtens, J.R. 1978. Adaptations and flood tolerance of tree species. In D.D. Hook and R.M.M. Crawford (eds.) *Plant Life in Anaerobic Environments*, pp. 299–331. Ann Arbor Science, Ann Arbor, Michigan.

Horton, J.S. 1964. Notes on the introduction of deciduous tamarisk. US Forest Service Research Note RM-50.

Horton, J.S. 1972. Management problems in phreatophyte and riparian zones. *Journal of Soil and Water Conservation* **27**: 57–61.

Horton, J.S. 1973. Evapotranspiration and watershed research as related to riparian and phreatophyte management, and abstract bibliography. USDA Miscellaneous Publication 1234.

Horton, J.S. 1977. The development and perpetuation of the permanent tamarisk type in the phreatophyte zone of the Southwest. In R.R. Johnson and D.A. Jones (eds.) *Importance, Preservation and Management of Riparian Habitat: A Symposium*. US Forest Service General Technical Report RM-43, pp. 124–7.

Horton, J.S., Mounts, F.C. and Kraft, J.M. 1960. Seed germination and seedling establishment of phreatophyte species. US Forest Service Research Paper RM-48.

Hosner, J.F. 1957. Effects of water upon the seed germination of bottomland trees. *Forest Science* **3**: 67–70.

Hosner, J.F. 1958. The effects of complete inundation upon seedlings of six bottomland tree species. *Ecology* **39**: 371–3.

Hosner, J.F. 1960. Relative tolerance to complete inundation of fourteen bottomland tree species. *Forest Science* **6**: 246–51.

Hosner, J.F. and Boyce, S.G. 1962. Tolerance to water saturated soil of various bottomland tree species. *Forest Science* **8**: 180–6.

Hosner, J.F. and Leaf, A.L. 1962. The effect of soil saturation upon the dry weight, ash content and nutrient absorption of various bottomland tree seedlings. *Soil Science Society Proceedings* **26**: 401–4.

Hosner, J.F., Leaf, A.L., Dickson, R. and Hart, J.B. 1965. Effects of varying soil moisture upon the nutrient uptake of four bottomland tree species. *Soil Science Society of America Journal* **29**: 313–16.

Hosner, J.F. and Minckler, L.S. 1963. Bottomland hardwood forests of southern Illinois – regeneration and succession. *Ecology* **44**: 29–41.

Howard, J.A. and Mitchell, C.W. 1985. *Phytogeomorphology*. Wiley, New York.

Howard, J.A. and Penfound, W.T. 1942. Vegetational studies in areas of sedimentation in the Bonnet Carre Floodway. *Bulletin of the Torrey Botanical Club* **69**: 281–9.

Howe, W.H. and Knopf, F.L. 1991. On the imminent decline of Rio Grande cottonwoods in central New Mexico. *Southwestern Naturalist* **36**: 218–24.

Huber, J. 1910. Mattase madeiros amazonicas. *Botanische Museum Goeldi* **6**: 91–225.

Huennecke, L.F. 1982. Wetland forests of Tompkins County, New York. *Bulletin of the Torrey Botanical Club* **109**: 51–63.

Huennecke, L.F. and Sharitz, R.R. 1990. Substrate heterogeneity and regeneration of a swamp tree, *Nyssa aquatica*. *American Journal of Botany* **77**: 413–19.

Huffman, R.T. 1980. The relation of flood timing and duration to variation in selected bottomland hardwood communities of southern Arkansas. Miscellaneous Paper EL-80-4. US Army Engineer Waterways Experiment Station, Vicksburg, Mississippi.

Huffman, R.T. and Forsythe, S.W. 1981. Bottomland hardwood forest communities and their relation to anaerobic soil conditions. In J.R. Clark and J. Benforado (eds.) *Wetlands of Bottomland Hardwood Forests*, pp. 187–96. Elsevier, Amsterdam.

Hughes, F.M.R. 1984. A comment on the impact of development schemes on the floodplain forests of the Tana River of Kenya. *Geographical Journal* **150**: 230–45.

Hughes, F.M.R. 1988. The ecology of African floodplain forests in semi-arid and arid zones: a review. *Journal of Biogeography* **15**: 127–40.

Hughes, F.M.R. 1990. The influence of flooding regimes on forest distribution and composition in the Tana River floodplain, Kenya. *Journal of Applied Ecology* **27**: 475–91.

Hukusima, T., Kershaw, K.A. and Takase, Y. 1986. The impact on the Senjogahara ecosystem of extreme run-off events from the River Sakasagawa, Nikko National park. I. Vegetation and its relationship to flood damage. *Ecological Research* **1**: 279–92.

Hunt, C.E. 1988. *Down By the River*. Island Press, Covelo, California.

Hupp, C.R. 1982. Stream-grade variation and riparian-forest ecology along Passage Creek, Virginia. *Bulletin of the Torrey Botanical Club* **109**: 488–99.

Hupp, C.R. 1983. Vegetation patterns on channel features in the Passage Creek gorge, Virginia. *Castanea* **48**: 62–72.

Hupp, C.R. 1986. The headward extent of fluvial landforms and associated vegetation on Massanutten Mountain, Virginia. *Earth Surface Processes and Landforms* **11**: 545–55.

Hupp, C.R. 1987. Channel widening and bank accretion determination through tree-ring analysis along modified West Tennessee streams. In G.C. Jacoby and J.W. Hornbeck (eds.) *Proceedings of the international symposium on ecological aspects of tree-ring research*. US Department of Energy CONF-8608144, pp. 224–33.

Hupp, C.R. 1988. Plant ecological aspects of flood geomorphology and paleoflood history. In V.R. Baker, R.C. Kochel and P.C. Patton (eds.) *Flood Geomorphology*, pp. 335–56. J. Wiley, New York.

Hupp, C.R. 1990. Vegetation patterns in relation to basin hydrogeomorphology. In J.B. Thornes (ed.) *Vegetation and Erosion*, pp. 217–37. Wiley, New York

Hupp, C.R. and Morris, E.E. 1990. A dendrogeomorphic approach to measurement of sedimentation in a forested wetland, Black Swamp, Arkansas. *Wetlands* **10**: 107–24.

Hupp, C.R. and Osterkamp, W.R. 1985. Bottomland vegetation distribution along Passage Creek, Virginia, in relation to fluvial landforms. *Ecology* **66**: 670–81.

Hupp, C.R. and Simon, A. 1986. Vegetation and bank-slope development. *Proceedings of the 4th Federal Interagency Sediment Conference*, vol. 2, Sec. 5, pp. 83–92.

Hupp, C.R. and Simon, A. 1991. Bank accretion and the development of vegetated depositional surfaces along modified alluvial channels. *Geomorphology* **4**: 111–24.

Huryn, A.D. and Wallace, J.B. 1987. Local geomorphology as a determinant of macrofaunal production in a mountain stream. *Ecology* **68**: 1932–42.

Huston, M., DeAngelis, D. and Post, W. 1988. New computer models unify ecological theory. *BioScience* **38**: 682–91.

Hutchinson, G.E. 1951. Copepodology for the ornithologist. *Ecology* **32**: 571–7.

Hynes, H.B.N. 1975. The stream and its valley. *Verhandlungen Internationale Vereiningung für Theoretische und Angewandte Limnologie* **19**: 1–15.

Ikeda, S. and Izumi, N. 1990. Width and depth of self-formed straight gravel rivers with bank vegetation. *Water Resources Research* **26**: 2353–64.

Illichevsky, S. 1933. The river as a factor of plant distribution. *Journal of Ecology* **21**, 436–41.

Irmler, U. 1977. Inundation forest types in the vicinity of Manaus. In P. Muller (ed.) *Ecosystem Research in South America, Biogeographica*, Volume 8, pp. 17–29. Junk, The Hague.

Irmler, U. and Furch, K. 1980. Weight, energy, and nutrient exchanges during the decomposition of leaves in the emersion phase of Central-Amazonian inundation forests. *Pedobiologia* **20**: 118–30.

Irvine, J.R. and West, N.E. 1979. Riparian tree species distribution and succession along the lower Escalante River, Utah. *Southwestern Naturalist* **24**: 331–46.

Jackson, J.R. and Lindauer, I.E. 1978. Vegetation of the flood plain of the South Platte River in the proposed Narrows Reservoir site. *Transactions, Missouri Academy of Science* **12**: 37–46.

Jackson, R.G. 1981. Sedimentology of muddy fine-grained channel deposits in meandering streams of the American Middle West. *Journal of Sedimentary Petrology* **51**: 1169–92.

Jacobs, T.C. and Gilliam, J.W. 1985*a*. Riparian losses of nitrate from agricultural drainage waters. *Journal of Environmental Quality* **14**: 472–8.

Jacobs, T.C. and Gilliam, J.W. 1985*b*. Headwater stream losses of nitrogen from two coastal plain watersheds. *Journal of Environmental Quality* **14**: 467–72.

James, C.S. 1985. Sediment transfer to overbank sections. *Journal of Hydraulic Research* **23**: 435–52.

James, C.S. 1987. The distribution of fine sediment deposits in compound channel systems. *Water SA* **13**: 7–14.

Jean, M. and Bouchard, A. 1991. Temporal changes in wetland landscapes of a section of the St. Lawrence River, Canada. *Environmental Management* **15**: 241–50.

Jemison, R.L. 1989. Conditions That Define a Riparian Zone in Southeastern

Arizona. Ph.D. dissertation, University of Arizona.

Jenkins, K.J. and Wright, R.G. 1987: Simulating succession of riparian spruce forests and white-tailed deer carrying capacity in northwestern Montana. *Western Journal of Applied Forestry* **2**: 80–3.

Jenkins, S.H. 1975. Food selection by beavers: a multidimensional contingency table analysis. *Oecologia* **21**: 157–73.

Jenkins, S.H. 1979. Seasonal and year-to-year differences in food selection by beavers. *Oecologia* **44**: 112–16.

Jenkins, S.H. 1980. A size – distance relation in food selection by beavers. *Ecology* **61**: 740–6.

Jepson, W.L. 1893. The riparian botany of the lower Sacramento. *Erythea* **1**: 230ff. (Cited by Thompson (1961, 1977).)

Johnson, F.L. and Bell, D.T. 1976*a*. Plant biomass and net primary production along a flood-frequency gradient in the streamside forest. *Castanea* **41**: 156–65.

Johnson, F.L. and Bell, D.T. 1976*b*. Tree growth and mortality in the streamside forest. *Castanea* **41**: 34–41.

Johnson, J.S. and Heede, W.B. 1975. Dispersal of Drosophila: the effect of baiting on the behavior and distribution of natural populations. *American Naturalist* **109**: 207–16.

Johnson, P.L. and Vogel, T.C. 1966. Vegetation of the Yukon Flats region, Alaska. US Army Cold Regions Research and Engineering Laboratory Research Report 209.

Johnson, R.R. 1971. Tree removal along southwestern rivers and effects on associated organisms. *American Philosophical Society Yearbook* **1970**: 321–2.

Johnson, R.R. and Haight, L.T. 1984. Riparian problems and initiatives in the American Southwest: a regional perspective. In R.E. Warner and K.M. Hendrix (eds.) *California Riparian Systems*, pp. 404–12. University of California Press, Berkeley.

Johnson, R.R. and Lowe, C.H. 1985. On the development of riparian ecology. In R.R. Johnson, C.D. Ziebell, D.R. Patton, P.F. Ffolliott, and R.H. Hamre (eds.) *Riparian ecosystems and their management: reconciling conflicting uses.* US Forest Service General Technical Report RM-120, pp. 112–16.

Johnson, R.R., Bennett, P.S. and Haight, L.T. 1989. Southwestern woody riparian vegetation and succession: an evolutionary approach. In D.L. Abell (ed.) *Proceedings of the California Riparian Systems Conference.* US Forest Service General Technical Report PSW-110, pp. 135–9.

Johnson, R.R., Carothers, S.W. and Simpson, J.M. 1984. A riparian classification system. In R.E. Warner and K.M. Hendrix (eds.) *California Riparian Systems*, pp. 375–82. University of California Press, Berkeley.

Johnson, W.C. and Brophy, J.A. 1982. Altered hydrology of the Missouri River and its effects on floodplain forest ecosystems. Virginia Water Resources Research Center Bulletin 139. Virginia Polytechnic and State University, Blacksburg.

Johnson, W.C., Burgess, R.L. and Keammerer, W.R. 1976. Forest overstory vegetation and environment on the Missouri River floodplain in North Dakota. *Ecological Monographs* **46**: 59–84.

Johnson, W.C., Reily, P.W., Andrews, L.S., McLellan, J.F. and Brophy, J.A. 1982. Altered hydrology of the Missouri River and its effects on floodplain forest ecosystems. Virginia Water Resources Research Center Bulletin 139. Virginia Polytechnic and State University, Blacksburg.

Johnson, W.C., Sharpe, D.M., DeAngelis, D.L., Fields, D.L. and Olson, R.J. 1981. Modeling seed dispersal and forest island dynamics. In R.L. Burgess and D.M. Sharpe (eds.) *Forest-Island Dynamics in Man-Dominated Landscapes*, pp. 215–39. Springer-Verlag, New York.

Johnston, C.A. and Naiman, R.J. 1987. Boundary dynamics at the aquatic-terrestrial interface: the influence of beaver and geomorphology. *Landscape Ecology* **1**: 47–57.

Johnston, C.A. and Naiman, R.J. 1990. Browse selection by beaver: effects on riparian forest composition. *Canadian Journal of Forest Research* **20**: 1036–43.

Johnston, C.A., Bubenzer, G.D., Lee, G.B., Madison, F.W. and McHenry, J.R. 1984. Nutrient trapping by sediment deposition in a seasonally flooded lakeside wetland. *Journal of Environmental Quality* **13**: 283–90.

Johnston, C.A., Detenbeck, N.E. and Niemi, G.J. 1990. The cumulative effect of wetlands on stream quality and quantity. A landscape approach. *Biogeochemistry* **10**: 105–41.

Jones, R.H. and Sharitz, R.R. 1989. Potential advantages and disadvantages of germinating early for trees in floodplain forests. *Oecologia* **81**: 443–9.

Jones, R.H., Sharitz, R.R. and McLeod, K.W. 1989. Effects of flooding and root competition on growth of shaded bottomland hardwood seedlings. *American Midland Naturalist* **121**: 165–75.

Julien, P.Y. and Frenette, M. 1986. LAVSED-II – A model for predicting suspended load in northern streams. *Canadian Journal of Civil Engineering* **13**: 162–70.

Junk, W.J., Bayley, P.B. and Sparks, R.E. 1989. The flood pulse concept in river-floodplain systems. In D.P. Dodge (ed.) *Proceedings of the International Large River Symposium. Canadian Special Publication in Fisheries and Aquatic Sciences* **106**: 110–27.

Kadlec, R.H. and Kadlec, J.A. 1978. Wetlands and water quality. In *Wetland Functions and Values: The State of Our Understanding*, pp. 436–56. American Water Resources Association, Bethesda, Maryland.

Kahn, F. and Mejia, K. 1990. Palm communities in wetland forest ecosystems of Peruvian Amazonia. *Forest Ecology and Management* **33/34**: 169–79.

Kalliola, R. and Puhakka, M. 1988. River dynamics and vegetation mosaicism: a case study of the River Kamajohka, northernmost Finland. *Journal of Biogeography* **15**: 703–19.

Karplus, W.J. 1976. The spectrum of mathematical modeling and systems simulation. In L. Dekker (ed.) *Simulation of Systems*. North Holland, Dordrecht, pp. 5–13.

Karr, J.R. and Schlosser, I.J. 1978. Water resources and the land-water interface. *Science* **201**: 229–34.

Katibah, E.F. 1984. A brief history of riparian forests in the Central Valley of California. In: R.E. Warner and K.M. Hendrix (eds.) *California Riparian Systems*, pp. 23–9. University of California Press, Berkeley.

Katibah, E.F., Nedeff, N.E. and Dummer, K.J. 1984. Summary of riparian vegetation areal and linear measurements from the Central Valley riparian mapping project. In: R.E. Warner and K.M. Hendrix (eds.) *California Riparian*

Systems, pp. 46–50. University of California Press, Berkeley.

Keammerer, W.R., Johnson, W.C. and Burgess, R.L. 1975. Floristic analysis of the Missouri River bottomland forests in North Dakota. *Canadian Field Naturalist* **89**: 5–19.

Keay, R.J.W. 1949. An example of Sudan zone vegetation in Nigeria. *Journal of Ecology* **37**: 335–64.

Keddy, P.A. 1984. Plant zonation on lakeshores in Nova Scotia: a test of the resource specialization hypothesis. *Journal of Ecology* **72**: 797–808.

Keddy, P.A. 1989. *Competition*. Routledge, Chapman and Hall, New York.

Keddy, P.A. and MacLellan, P. 1990. Centrifugal organization in forests. *Oikos* **59**: 75–84.

Keeley, J.E. and Zedler, P.H. 1978. Reproduction of chaparral shrubs after fire: a comparison of sprouting and seeding strategies. *American Midland Naturalist* **99**: 142–61.

Keller, E.A. and Melhorn, W.N. 1973. Bedforms and fluvial processes in alluvial stream channels: selected observations. In M. Morisawa (ed.) *Fluvial Geomorphology*, pp. 253–84. State University of New York, Binghamton, New York.

Keller, E.A. and Swanson, F.J. 1979. Effects of large organic material on channel form and fluvial processes. *Earth Surface Processes* **4**: 361–80.

Keller, E.A. and Tally, T. 1979. Effects of large organic debris on channel form and fluvial processes in the coastal redwood environment. In D.D. Rhodes and G.P. Williams (eds.) *Adjustments of the Fluvial System*, pp. 169–97. Kendall/Hunt, Dubuque, Iowa.

Kellman, M.C. 1970. The influence of accessibility on the composition of vegetation. *Professional Geographer* **22**: 1–4.

Kellogg, R.S. 1905. Forest belts of western Kansas and Nebraska. US Forest Service Bulletin No. 66.

Kelsey, H.M., Lamberson, R. and Madej, M.A. 1987. Stochastic model for the long-term transport of stored sediment in a river channel. *Water Resources Research* **23**: 1738–50.

Kennedy, H.E. and Krinard, R.M. 1974. Mississippi River flood's impact on natural hardwood forests and plantations. US Forest Service Research Note SO-177.

Kesel, R.H. 1988. The decline in the suspended load of the lower Mississippi river and its influence on adjacent wetlands. *Environmental Geology and Water Science* **11**: 271–81.

Kesel, R.H. 1989. The role of the Mississippi river in wetland loss in southeastern Louisiana. *Environmental Geology and Water Science* **13**: 183–93.

Kessell, S.R. 1979. *Gradient Modeling*. Springer-Verlag, New York.

Killingbeck, K.T. 1986. Litterfall dynamics and element use efficiency in a Kansas gallery forest. *American Midland Naturalist* **116**: 180–9.

Killingbeck, K.T. and Wali, M.K. 1978. Analysis of a North Dakota gallery forest: nutrient, trace element, and productivity relations. *Oikos* **30**: 29–60.

Kindschy, R.R. 1985. Response of red willow to beaver use in southeastern Oregon. *Journal of Wildlife Management* **49**: 26–8.

King, A.W. 1991. Translating models across scales in the landscape. In M.G. Turner and R.H. Gardner (eds.) *Quantitative Methods in Landscape Ecology*, pp. 479–517. Springer-Verlag, New York.

Kleiss, B.A., Morris, E.E., Nix, J.F. and Barko, J.W. 1989. Modification of riverine water quality by an adjacent bottomland hardwood wetland. In D.W. Fisk (ed.) *Wetlands: Concerns and Successes*, pp. 429–38. American Water Resources Association, Bethesda, Maryland.

Klimas, C.V., Martin, C.O. and Teaford, J.W. 1981. Impacts of flooding regime modification on wildlife habitats of bottomland hardwood forests in the lower Mississippi Valley. Technical Report EL-81–13. US Army Engineers Waterways Experiment Station, Vicksburg, Mississippi.

Klimo, E. and Prax, A. 1985. Soil conditions. In M. Penka, M. Vyskot, E. Klimo and F. Vasicek (eds.) *Floodplain Forest Ecosystem. I. Before Water Management Measures*, pp. 61–78. Elsevier, Amsterdam.

Klinge, H. and Furch, K. 1991. Towards the classification of Amazonian floodplains and their forests by means of biogeochemical criteria of river water and forest biomass. *Interciencia* **16**: 196–201.

Klinge, H., Junk, W.J. and Revilla, C.J. 1990. Status and distribution of forested wetlands in tropical South America. *Forest Ecology and Management* **33/34**: 81–101.

Klock, G.O. 1985. Modeling the cumulative effects of forest practices on downstream aquatic ecosystems. *Journal of Soil and Water Conservation* **40**, 237–41.

Klokk, T. 1981. Classification and ordination of river bank vegetation from middle and upper parts of the River Gaula, central Norway. *Kongelige Norske Videnskabers Selskab Skrifter* **2**: 1–43.

Klopatek, J.M. 1978. Nutrient dynamics of freshwater riverine marshes and the role of emergent macrophytes. In R.E. Good, D.F. Whigham and R.L. Simpson (eds.) *Freshwater Wetlands*, pp. 195–216. Academic Press, New York.

Knisel, W.G. (ed.) 1980. CREAMS: a field-scale model for chemicals, runoff, and erosion from agricultural management systems. Conservation Research Report No. 26. US Department of Agriculture, Washington.

Knopf, F.L. 1985. Significance of riparian vegetation to breeding birds across an altitudinal cline. In R.R. Johnson, C.D. Ziebell, D.R. Patton, P.F. Ffolliott and R.H. Hamre (eds.) *Riparian Ecosystems and Their Management: Reconciling Conflicting Uses*. US Forest Service General Technical Report RM-120, pp. 105–10.

Knopf, F.L. 1986. Changing landscapes and the cosmopolitanism of the eastern Colorado avifauna. *Wildlife Society Bulletin* **14**: 132–42.

Knopf, F.L. 1988. Guild structure of a riparian avifauna relative to seasonal cattle grazing. *Journal of Wildlife Management* **52**: 280–90.

Knopf, F.L. 1989. Riparian wildlife habitats: more, worth less, and under invasion. In *Restoration, Creation, and Management of Wetland and Riparian Ecosystems in the American West*, pp. 20–2. Rocky Mountain Chapter, Society of Wetland Scientists, Boulder, Colorado.

Knopf, F.L. and Olson, T.E. 1984. Naturalization of Russian-olive: implications to Rocky Mountain wildlife. *Wildlife Society Bulletin* **12**: 289–98.

Knopf, F.L. and Scott, M.L. 1990. Altered flows and created landscapes in the Platte River headwaters, 1840–1990. In J.M. Sweeney (ed.) *Management of Dynamic Ecosystems*, pp. 47–70. North Central Section, The Wildlife Society, West Lafayette, Indiana.

Knopf, F.L., Johnson, R.R., Rich, T., Samson, F.B. and Szaro, R.C. 1988.

Conservation of riparian ecosystems in the United States. *Wilson Bulletin* **100**: 272–84.

Knox, J.C. 1977. Human impacts on Wisconsin stream channels. *Annals of the Association of American Geographers* **67**: 323–42.

Knox, J.C. 1987. Historical valley floor sedimentation in the upper Mississippi valley. *Annals of the Association of American Geographers* **77**: 224–44.

Knudsen, J., Thomsen, A. and Refsgaard, J.C. 1986. WATBAL, a semi-distributed, physically based hydrological modelling system. *Nordic Hydrology* **17**: 347–62.

Kochel, R.C., Ritter, D.F. and Miller, J. 1987. Role of tree dams in the construction of pseudo-terraces and variable geomorphic response to floods in Little River valley, Virginia. *Geology* **15**: 718–21.

Kondolf, G.M., Webb, J.W., Sale, M.J. and Felando, T. 1987. Basic hydrological studies for assessing impacts of flow diversions on riparian vegetation: examples from streams of the eastern Sierra Nevada, California, USA. *Environmental Management* **11**: 757–69.

Kovalchik, B.L. and Chitwood, L.A. 1990. Use of geomorphology in the classification of riparian plant associations in mountainous landscapes of central Oregon, USA. *Forest Ecology and Management* **33/34**: 405–18.

Kozlowski, T.T. 1984*a*. Plant responses to flooding of soil. *BioScience* 34, 162–7.

Kozlowski, T.T. 1984*b*. Responses of woody plants to flooding. In T.T. Kozlowski (ed.) *Flooding and Plant Growth*, pp. 129–63. Academic Press, Orlando.

Kramer, P.J. 1951. Causes of injury to plants resulting from flooding of the soil. *Plant Physiology* **26**: 727–36.

Krishnappan, B.G. and Lau, Y.L. 1986. Turbulence modeling of flood plain flows. *Journal of Hydraulic Engineering* **112**: 251–66.

Kubitzki, K. 1989. The ecogeographical differentiation of Amazonian inundation forests. *Plant Systematics and Evolution* **162**: 285–304.

Kuchler, A.W. 1964. *Potential Natural Vegetation of the Coterminous United States.* American Geographical Society Special Publication No. 36, Washington.

Kuhlman, M. and Kuhn, E. (1947). *A Flora do Distrito de Ibiti.* Publication of the Institute of Botany, Secretariat of Agriculture, Sao Paulo.

Kupfer, J.A. 1990*a*. Effects of connectivity and distance on riparian tree community internal structure and species richness. Unpublished manuscript, Department of Geography, University of Iowa.

Kupfer, J.A. 1990*b*. Modeling primary and secondary effects of climatic change on floodplain vegetation dynamics. Unpublished manuscript, Department of Geography, University of Iowa.

Kupfer, J.A. and Malanson, G.P. 1992*a*. Nonequilibrium structure and composition of a riparian forest edge. *American Midland Naturalist* (submitted).

Kupfer, J.A. and Malanson, G.P. 1992*b*. Directional change in riparian forest composition at an eroding cutbank. *Landscape Ecology*, in press.

Laing, D. and Stockton, C.W. 1976. Riparian dendrochronology: a method for determining flood histories of ungaged watersheds. National Technical Information Service PB 256 967.

LaMarche, V.C. 1966. An 800-year history of stream erosion as indicated by botanical evidence. US Geological Survey Professional Paper 550-D.

Lambert, J.M. 1951. Alluvial stratigraphy and vegetational succession of the Bure

Valley Broads. III. Classification status and distribution of communities. *Journal of Ecology* **139**: 149–70.

Lamotte, S. 1990. Fluvial dynamics and succession in the Lower Ucayali River basin, Peruvian Amazonia. *Forest Ecology and Management* **33/34**: 141–56.

Landwehr, J.M. and Matalas, N.C. 1986. On the nature of persistence in dendrochronologic records with implications for hydrology. *Journal of Hydrology* **86**: 239–77.

Langdon, O.G., McClure, J.P., Hook, D.D., Crockett, J.M. and Hunt, R. 1981. Extent, condition, management, and research needs of bottomland hardwood-cypress forests in the Southeast. In J.R. Clark and J. Benforado (eds.) *Wetlands of Bottomland Hardwood Forests*, pp. 71–86. Elsevier, Amsterdam.

Langley, R.D. 1984. SOFAR: a small-town water diversion project on the South Fork, American River. In R.E. Warner and K.M. Hendrix (eds.) *California Riparian Systems*, pp. 58–67. University of California Press, Berkeley.

Lant, C.L. and Tobin, G.A. 1989. The economic value of riparian corridors on Cornbelt floodplains: a research framework. *Professional Geographer* **41**: 337–49.

Lant, C.L. and Roberts, R.S. 1990. Greenbelts in the Cornbelt: riparian wetlands, intrinsic values, and market failure. *Environment and Planning* A**22**: 1375–88.

Lawton, R.M. 1967. The conservation and management of the riparian evergreen forests of Zambia. *Commonwealth Forestry Review* **46**: 223–32.

Laymon, S.A. 1984. Photodocumentation of vegetation and landform change on a riparian site, 1880–1980: Dog Island, Red Bluff, California. In R.E. Warner and K.M. Hendrix (eds.) *California Riparian Systems* pp. 150–8. University of California Press, Berkeley.

Lee, L.C. 1979. A Study of Plant Associations in Upland Riparian Habitats in Western Montana. M.S. thesis, University of Montana, Missoula.

Lee, L.C. 1983. The Floodplain and Wetland Vegetation of Two Pacific Northwest River Ecosystems. Ph.D. dissertation, University of Washington.

Lee, L.C. and Hinckley, T.M. 1982. Impact of water level changes on woody riparian and wetland communities. Vol. IX: The Alaska region. US Fish and Wildlife Service OBS-82/22.

Lee, L.C., Hinckley, T.M. and Scott, M.L. 1985. Plant water status relationships among major floodplain sites of the Flathead River, Montana. *Wetlands* **5**: 15–34.

Lee, M.B. 1945. An ecological study of the floodplain forest along the White River system in Indiana. *Butler University Botanical Studies* **7**: 155–65.

Leighton, J.P. and Risser, R.J. 1989. A riparian vegetation ecophysiological response model. In D.L. Abell (ed.) *Proceedings of the California Riparian Systems Conference*. US Forest Service General Technical Report PSW-110, pp. 370–4.

Leitman, H.M., Sohm, J.E. and Franklin, M.A. 1983. Wetland hydrology and tree distribution of the Apalachicola River flood plain, Florida. US Geological Survey Water Supply Paper 2186-A.

Leitman, H.M., Darst, M.R. and Nordhaus, J.J. 1991. Fishes in the forested flood plain of the Ochlockonee River, Florida, during flood and drought conditions. US Geological Survey Water-Resources Investigations Report 90–4204.

Leonard, R.M. 1988. Integrating riparian habitat management objectives with livestock grazing on National Forest system lands. Ph.D. dissertation, Texas Tech University, Lubbock.

Leopold, L.B., Wolman, M.G. and Miller, J.P. 1964. *Fluvial Processes in Geomorphology*. W.H. Freeman, San Francisco.

Lesak, J. 1985. Primary production of a grassland ecosystem of floodplain meadows. In M. Penka, M. Vyskot, E. Klimo, and F. Vasicek (eds.) *Floodplain Forest Ecosystem. I. Before Water Management Measures*, pp. 275–88. Elsevier, Amsterdam.

Levin, S.A. 1976. Population dynamic models in heterogeneous environments. *Annual Review of Ecology and Systematics* **7**: 287–310.

Levin, S.A. 1991. Spatial phenomena in ecological systems: some introductory remarks. *Bulletin of the Ecological Society of America* **72**(2): 173.

Li, R.M. and Shen, H.S. 1973. Effect of tall vegetation on flow and sediment. *Journal of the Hydraulics Division, American Society of Civil Engineers* (ASCE) HY**5**: 793–814.

Licht, L.A. 1990. Poplar Tree Buffer Strips Grown in Riparian Zones. Ph.D. dissertation, University of Iowa, Iowa City.

Licht, L.A. and Schnoor, J.L. 1989. Poplar tree buffer strips grown in riparian zones for biomass production and non-point source pollution control. 1988–89 Final Report to the Leopold Center for Sustainable Agriculture, Ames, Iowa.

Lienkaemper, G.W. and Swanson, F.J. 1987. Dynamics of large woody debris in streams of old-growth Douglas-fir forests. *Canadian Journal of Forest Research* **17**: 150–6.

Likens, G.E. and Bormann, F.H. 1974. Linkages between terrestrial and aquatic ecosystems. *BioScience* **24**: 447–56

Lindauer, I.E. 1983. A comparison of the plant communities of the South Platte and Arkansas River drainages in eastern Colorado. *Southwestern Naturalist* **28**: 249–59.

Lindsey, A.A., Petty, R.O., Sterling, D.K. and VanAsdall, W. 1961. Vegetation and environment along the Wabash and Tippecano rivers. *Ecological Monographs* **31**: 105–56.

Lisle, T.E. 1989. Channel-dynamic control on the establishment of trees after large floods in northwestern California. In D.L. Abell (ed.) *Proceedings of the California Riparian Systems Conference*. US Forest Service General Technical Report PSW-110, pp. 9–13.

Little, E.L. 1971. *Atlas of United States Trees*. Vol. 1. *Conifers and Important Hardwoods*. US Forest Service Misc. Publ. 1146.

Little, E.L. 1977. *Atlas of United States Trees*. Vol. 4. *Minor Eastern Hardwoods*. US Forest Service Misc. Publ. 1342.

Liu, Z.-J. and Malanson, G.P. 1991. Long-term cyclic dynamics of simulated riparian forest stands. *Forest Ecology and Management* **48**: 217–31.

Loucks, W.L. and Keen, R.A. 1973. Submersion tolerance of selected seedling trees. *Journal of Forestry* **71**: 496–7.

Lowrance, R. and Shirohammadi, A. 1985. REM: a model for Riparian Ecosystem Management in agricultural watersheds. In R.R. Johnson, C.D. Ziebell, D.R. Patton, P.F. Ffolliott and R.H. Hamre (eds.) *Riparian Ecosystems and Their Management: Reconciling Conflicting Uses*. US Forest Service General Technical Report RM-120, pp. 237–40.

Lowrance, R., Leonard, R. and Sheridan, J. 1985. Managing riparian ecosystems to control nonpoint pollution. *Journal of Soil and Water Conservation* **40**: 87–91.

Lowrance, R., McIntyre, S. and Lance, C. 1988. Erosion and deposition in a field forest system estimated using Cesium-137 activity. *Journal of Soil and Water Conservation* **43**: 195–9.

Lowrance, R., Sharpe, J.K. and Sheridan, J.M. 1986. Long-term sediment deposition in the riparian zone of a coastal plain watershed. *Journal of Soil and Water Conservation* **41**, 266–71.

Lowrance, R.R., Todd, R.L. and Asmussen, L.E. 1983. Waterborne nutrient budgets for the riparian zone of an agricultural watershed. *Agriculture, Ecosystems and Environment* **10**: 371–84.

Lowrance, R.R., Todd, R.L. and Asmussen, L.E. 1984*a*. Nutrient cycling in an agricultural watershed: I. Phreatic movement. *Journal of Environmental Quality* **13**: 22–6.

Lowrance, R.R., Todd, R.L. and Asmussen, L.E. 1984*b*. Nutrient cycling in an agricultural watershed: II. Streamflow and artificial drainage. *Journal of Environmental Quality* **13**: 27–32.

Lowrance, R., Todd, R., Fail, J., Hendrickson, O., Leonard, R. and Asmussen, L. 1984*c*. Riparian forests as nutrient filters in agricultural watersheds. *BioScience* **34**: 374–77.

Luftensteiner, H.W. 1979. The eco-sociological value of dispersal spectra of two plant communities. *Vegetatio* **41**: 61–7.

Lugo, A.E., Brinson, M. and Brown, S. eds. 1990. *Forested Wetlands. Ecosystems of the World* 15. Elsevier, Amsterdam.

MacArthur, R.H. 1972. *Geographical Ecology*. Princeton University Press, Princeton.

MacArthur, R.H. and Wilson, E.O. 1967. *The Theory of Island Biogeography*. Princeton University Press, Princeton.

Macbride, T.H. 1896. Notes on the forest distribution of Iowa. *Iowa Academy of Sciences* **3**: 96–101.

Madgwick, J. 1988. Somalia Research Project, and ecological study of the remaining areas of riverine forest in the Jubba Valley, southern Somalia. *Bulletin of the British Ecological Society* **19**: 19–24.

Magette, W.L., Brinsfield, R.B., Palmer, R.E. and Wood, J.D. 1989. Nutrient and sediment removal by vegetated filter strips. *Transactions of the American Society of Agricultural Engineers* (ASAE) **32**: 663–7.

Maher, L.J. 1982. The palynology of Devils Lake, Sauk County, Wisconsin. In *Quaternary History of the Driftless Area*. University of Wisconsin Extension, Geological and Natural History Survey, Field Trip Guide Book No. 5, pp. 119–35.

Malanson, G.P. 1980. Habitat and plant distributions in hanging gardens of the Narrows, Zion National Park, Utah. *Great Basin Naturalist* **40**: 178–82.

Malanson, G.P. 1982. The assembly of hanging gardens: effects of age, area, and location. *American Naturalist* **119**: 145–50.

Malanson, G.P. 1984. Intensity as a third factor of disturbance regime and its effect on species diversity. *Oikos* **43**: 411–13.

Malanson, G.P. 1985*a*. Simulation of competition between alternative shrub life history strategies through recurrent fires. *Ecological Modelling* **27**: 271–83

Malanson, G.P. 1985*b*. Spatial autocorrelation and distributions of plant species on environmental gradients. *Oikos* **45**: 278–80.

Malanson, G.P. 1987. Diversity, stability, and resilience: effects of fire regime. In L. Trabaud (ed.) *Role of Fire in Ecological Systems*, pp. 49–63. SPB Academic, The Hague.

Malanson, G.P. 1992. Disturbance intensity and spatial pattern and scale in the ecology of fragmented natural landscapes. *Ekistics*, in press.

Malanson, G.P. and Armstrong, M.P. 1990. Improving environmental simulation models to assess climate change impacts. University of Iowa, Department of Geography Discussion Paper No. 43.

Malanson, G.P. and Butler, D.R. 1990. Woody debris, sediment, and riparian vegetation of a subalpine river, Montana, USA. *Arctic and Alpine Research* **22**: 183–94.

Malanson, G.P. and Butler, D.R. 1991. Floristic variation among gravel bars in a subalpine river. *Arctic and Alpine Research* **23**: 273–8.

Malanson, G.P. and Butler, D.R. 1992. Tree-ring responses at a cutbank edge along a subalpine river. *Physical Geography* (submitted).

Malanson, G.P. and Kay, J. 1980. Flood frequency and the assemblage of dispersal types in hanging gardens of the Narrows, Zion National Park, Utah. *Great Basin Naturalist* **40**: 365–71.

Malanson, G.P. and Kupfer, J.A. 1993. Simulated fate of leaf litter and woody debris at a riparian cutbank. *Canadian Journal of Forest Research*, in press.

Malanson, G.P., Butler, D.R. and Walsh, S.J. 1990. Chaos theory in physical geography. *Physical Geography* **11**: 293–304.

Malanson, G.P., Butler, D.R. and Georgakakos, K.P. 1992. Nonequilibrium geomorphic processes and deterministic chaos. *Geomorphology* **5**: 311–22.

Malecki, R.A., Lassoie, J.R., Rieger, E. and Seamans, T. 1983. Effects of long-term artificial flooding on a northern bottomland hardwood forest community. *Forest Science* **29**: 535–44.

Mander, U. 1991. Eco-engineering methods to control nutrient losses from agricultural watersheds. In *European Seminar on Practical Landscape Ecology*, pp. 52–64. Roskilde Universitetsforlag GeoRuc, Roskilde.

Mander, U.E., Metsur, M.O. and Kulvik, M.E. 1989. Storungen des Stoffkreislaufs, des Energieflusses und des Bios als Kriterien für die Bestimmung der Belastung der Landschaft. *Petermanns Geographische Mitteilungen* **133**: 233–44.

Mark, D.M. 1975. Computer analysis of topography: a comparison of terrain storage methods. *Geografiska Annaler* **57**A: 179–88.

Marston, R.A. 1982. The geomorphic significance of log steps in forest streams. *Annals of the Association of American Geographers* **72**: 99–108.

Marston, R.A. 1991. Channel morphology and dynamics as applied to riparian systems. In S. Anderson, W. Hubert and F. Lindzey (eds.) *Proceedings of the Riparian Workshop*, pp. 15–23. Wyoming Cooperative Fish and Wildlife Research Unit, Laramie.

Marston, R.A. 1992. *Changes in Geomorphic Processes in the Snake River Following Impoundment of Jackson Lake and Potential Changes Due to 1988 Fires in the Watershed*. Unpublished Final Report to the National Park Service, Rocky Mountain Regional Office, Denver, Colorado.

Marston, R.A. and Anderson, J.E. 1991. Watersheds and vegetation of the Greater

Yellowstone Ecosystem. *Conservation Biology* **5**: 338–46.

Martin, C.W. and Johnson, W.C. 1987. Historical channel narrowing and riparian vegetation expansion in the Medicine Lodge River basin, Kansas, 1871–1983. *Annals of the Association of American Geographers* **77**: 436–49.

Mason, C.F. and MacDonald, S.M. 1990. The riparian woody plant community of regulated rivers in eastern England. *Regulated Rivers* **5**: 159–66.

Mather, J.R., Field, R.T., Kalkstein, L.S. and Willmott, C.J. 1980. Climatology: the challenge for the eighties. *Professional Geographer* **32**: 285–92.

Mathews, C.P. and Kowalczewski, A. 1969. The disappearance of leaf litter and its contribution to production in the River Thames. *Journal of Ecology* **57**: 543–52.

May, R.M. 1976. Simple mathematical models with very complicated dynamics. *Nature* **261**: 459–67.

Mayack, D.T., Thorp, J.H. and Cothran, M. 1989. Effects of burial and floodplain retention on stream processing of allochthonous litter. *Oikos* **54**: 378–88.

McAlpine, R.G. 1961. Yellow-poplar seedlings intolerant to flooding. *Journal of Forestry* **59**: 566–8.

McArthur, J.V. and Marzolf, G.R. 1986. Interactions of the bacterial assemblages of a prairie stream with dissolved organic carbon from riparian vegetation. *Hydrobiologia* **134**: 193–9.

McArthur, J.V. and Marzolf, G.R. 1987. Changes in soluble nutrients of prairie riparian vegetation during decomposition on a floodplain. *American Midland Naturalist* **117**: 26–34.

McBride, J. 1973. Natural replacement of disease-killed elms. *American Midland Naturalist* **90**: 300–6.

McBride, J.R. and Strahan, J. 1984*a*. Fluvial processes and woodland succession along Dry Creek, Sonoma County, California. In R.E. Warner and K.M. Hendrix (eds.) *California Riparian Systems*, pp. 110–19. University of California Press, Berkeley.

McBride, J.R. and Strahan, J. 1984*b*. Establishment and survival of woody riparian species on gravel bars of an intermittent stream. *American Midland Naturalist* **112**: 235–45.

McConnaughay, K.D.M. and Bazzaz, F.A. 1991. Is physical space a soil resource? Ecology **72**: 94–103.

McDermott, R.E. 1954. Effects of saturated soil on seedling growth of some bottomland hardwood species. *Ecology* **35**: 36–41.

McIntyre, R. 1981. Death of a beaver pond. *Alaska* **47**: 50.

McCullough, J.D. 1988. Physicochemical characteristics of nine first-order streams under three riparian management regimes in east Texas. M.S. thesis, Stephen F. Austin State University, Nacogdoches, TX.

McKnight, J.S., Hook, D.D., Langdon, O.G. and Johnson, R.L. 1981. Flood tolerance and related characteristics of trees of the bottomland forests of the southern United States. In J.R. Clark and J. Benforado (eds.) *Wetland of Bottomland Hardwood Forests*, pp. 29–70. Elsevier, Amsterdam.

McLellan, B.N. 1989. Effects of Resource Extraction Industries on Behaviour and Population Dynamics of Grizzly Bears in the Flathead Drainage, British Columbia and Montana. Ph.D. dissertation, University of British Columbia, Vancouver.

McLeod, K.W. and McPherson, J.K. 1973. Factors limiting the distribution of *Salix nigra*. *Bulletin of the Torrey Botanical Club* **100**: 102–10.

McVaugh, R. 1947. Establishment of vegetation on sandflats along the Hudson River. *Ecology* **28**: 189–94.

Meave, J., Kellman, M., MacDougall, A. and Rosales, J. 1991. Riparian habitats as tropical forest refugia. *Global Ecology and Biogeography Letters* **1**: 69–76.

Medina, A.L. 1986. Riparian plant communities of the Fort Bayard watershed in southern New Mexico. *Southwestern Naturalist* **31**: 345–59.

Melanson, J.C. 1991. Multiobjective river corridor management – an invited comment. *Natural Hazards Observer* **16**(2):1–3.

Melick, D.R. 1990*a*. Relative drought resistance of *Tristaniopsis laurina* and *Acmena smithii* from riparian warm temperate rainforest in Victoria. *Australian Journal of Botany* **38**: 361–70.

Melick, D.R. 1990*b*. Flood resistance of *Tristaniopsis laurina* and *Acmena smithii* from riparian warm temperate rainforest in Victoria. *Australian Journal of Botany* **38**: 371–81.

Menges, E.S. 1986. Environmental correlates of herb species composition in five southern Wisconsin floodplain forests. *American Midland Naturalist* **115**: 106–17.

Menges, E.S. and Waller, D.M. 1983. Plant strategies in relation to elevation and light in floodplain herbs. *American Naturalist* **122**: 454–73.

Metzler, K.J. and Damman, A.W.H. 1985. Vegetation patterns in the Connecticut River floodplain in relation to frequency and duration of flooding. *Nature Canada* **112**: 535–47.

Meyer, J.L. 1990. A blackwater perspective on riverine ecosystems. *BioScience* **40**: 643–51.

Meyer, P.A. 1984. Economic and social values in riparian systems. In R.E. Warner and K.M. Hendrix (eds.) *California Riparian Systems*, pp. 216–20. University of California Press, Berkeley.

Meyer, P.A. 1985. A summary of socio-economic presentations at the first north American conference on riparian ecosystems and their management. In R.R. Johnson, C.D. Ziebell, D.R. Patton, P.F. Ffolliott and R.H. Hamre (eds.) *Riparian Ecosystems and Their Management: Reconciling Conflicting Uses*. US Forest Service General Technical Report RM-120, pp. 1–2.

Miller, T.B. and Johnson, F.D. 1977. Ecology of riparian communities dominated by white alder in western Idaho. In *Proceedings of the Symposium on Terrestrial and Aquatic Studies of the Northwest*, pp. 111–23. Eastern Washington State College Press, Cheney.

Milne, B.T. 1991. Lessons from applying fractal models to landscape patterns. In M.G. Turner and R.H. Gardner (eds.) *Quantitative Methods in Landscape Ecology*, pp. 199–235. Springer-Verlag, New York.

Mitsch, W.J., Dorge, G.L. and Wiemhoff, J.R. 1979*a*. Ecosystem dynamics and a phosphorous budget of an alluvial swamp in southern Illinois. *Ecology* **60**: 1116–24.

Mitsch, W.J., Rust, W., Behnke, A. and Lai, L. 1979*b*. Environmental Observations of a Riparian Ecosystem During Flood Season. Project No. A-099-ILL, Illinois Water Resources Center, Urbana.

Mitsch, W.J., Taylor, J.R. and Benson, K.B. 1991. Estimating primary productivity of forested wetland communities in different hydrologic landscapes. *Landscape Ecology* **5**: 75–92.

Molenaar, C.M. 1985. *Field and River Trip Guide to Canyonlands Country*. Society of Economic Paleontologists and Mineralogists, Golden, Colorado.

Mollitor, A.V., Leaf, A.L. and Morris, L.A. 1980. Forest soil variability on northeastern flood plains. *Soil Science Society of America Journal* **44**: 617–20.

Mooney, H.A. and Drake, J.A. 1986. *Ecology of Biological Invasions of North America and Hawaii*. Springer Verlag, New York.

Moore, I.D. and Burch, G.J. 1986. Modeling erosion and deposition: topographic effects. *Transactions of the American Society of Agricultural Engineers (ASAE)* **29**: 1624–40.

Moore, I.D., Grayson, R.B. and Ladson, A.R. 1991. Digital terrain modelling: a review of hydrological, geomorphological, and biological applications. *Hydrological Processes* **5**: 3–30.

Morin, E., Bouchard, A. and Jutras, P. 1989. Ecological analysis of disturbed riverbanks in the Montreal area of Quebec. *Environmental Management* **13**: 215–25.

Morisawa, M. 1985. *Rivers*. Longman, London.

Mosley, M.P. 1981. The influence of organic debris on channel morphology and bedload transport in a New Zealand forest stream. *Earth Surface Processes and Landforms* **6**: 572–9.

Mulholland, P.J. 1981. Organic carbon flow in a swamp-stream ecosystem. *Ecological Monographs* **51**: 307–22.

Munro, D.S. 1979. Daytime energy exchange and evaporation from a wooded swamp. *Water Resources Research* **15**: 1259–65.

Murphey, J.B. and Grissinger, E.H. 1985. Channel cross-section changes in Mississippi's Goodwin Creek. *Journal of Soil and Water Conservation* **40**: 148–53.

Murray, D.R. 1986. Seed dispersal by water. In D.R. Murray (ed.) *Seed Dispersal*, pp. 49–85. Academic Press, Sydney.

Muzika, R.M., Gladden, J.B. and Haddock, J.D. 1987. Structural and functional aspects of succession in southeastern floodplain forests following major disturbance. *American Midland Naturalist* **117**: 1–9.

Myers, R.L. 1990. Palm swamps. In A.E. Lugo, M.Brinson and S. Brown (eds.) *Forested Wetlands. Ecosystems of the World* 15, pp. 267–86. Elsevier, Amsterdam.

Myster, R.W. and McCarthy, B.C. 1989. Effects of herbivory and competition on survival of *Carya tomentosa* (Juglandaceae) seedlings. *Oikos* **56**: 145–8.

Nachlinger, J.L., Fox, C.A. and Moen, P.A. 1989*a*. Riparian vegetation base-line analysis and monitoring along Bishop Creek, California. In D.L. Abell (ed.) *Proceedings of the California Riparian Systems Conference*. US Forest Service General Technical Report PSW-110, pp. 387–92.

Nachlinger, J.L., Smith, S.D. and Risser, R.J. 1989*b*. Riparian plant water relations along the North Fork, Kings River, California. In D.L. Abell (ed.) *Proceedings of the California Riparian Systems Conference*. US Forest Service General Technical Report PSW-110, pp. 366–9.

Naiman, R.J. and Decamps, H. (eds.) 1990. *The Ecology and Management of Aquatic-Terrestrial Ecotones. Man and the Biosphere Series* Vol. 4. UNESCO, Paris.

Naiman, R.J. and Melillo, J.M. 1984. Nitrogen budget of a subarctic stream altered by beaver (*Castor canadensis*). *Oecologia* **62**: 150–5.

Naiman, R.J., Decamps, H., Pastor, J. and Johnston, C.A. 1988*a*. The potential importance of boundaries to fluvial ecosystems. *Journal of the North American Benthological Society* **7**: 289–306.

Naiman, R.J., Holland, M.M., Decamps, H. and Risser, P.G. 1988*b*. A new UNESCO Programme: research and management of land/inland water ecotones. Biology International, Special Issue **17**: 107–36.

Naiman, R.J., Johnston, C.A. and Kelley, J.C. 1988*c*. Alteration of North American streams by beaver. *BioScience* **38**: 753–62.

Naiman, R.J., Manning, T. and Johnston, C.A. 1991. Beaver population fluctuations and tropospheric methane emissions in boreal wetlands. *Biogeochemistry* **12**: 1–15.

Naiman, R.J., Melillo, J.M. and Hobbie, J.E. 1986. Ecosystem alteration of boreal forest streams by beaver (Castor canadensis). *Ecology* **67**: 1254–69.

Naiman, R.J., Melillo, J.M., Lock, M.A., Ford, T.E. and Reice, S.R. 1987. Longitudinal patterns of ecosystem processes and community structure in a subarctic river continuum. *Ecology* **68**: 1139–56.

Nanson, G.C. 1980. Point bar and floodplain formation of the meandering Beatton River, northeastern British Columbia. *Sedimentology* **27**: 3–29.

Nanson, G.C. and Beach, H.F. 1977. Forest succession and sedimentation on a meandering-river floodplain, northeast British Columbia, Canada. *Journal of Biogeography* **4**: 229–51.

Nanson, G.C. and Hickin, E.J. 1983. Channel migration and incision on the Beatton River. *Journal of Hydraulic Engineering* **109**: 327–37.

Naveh, Z. and Lieberman, A.S. 1984. *Landscape Ecology; Theory and Application*. Springer-Verlag, New York.

Neef, E. 1963. Topologische und chorologische Arbeitsweisen in der Landschaftsforschung. *Petermanns Geographische Mitteilungen* **107**: 2349–59.

Neiff, J.J. and Poi de Neiff, A. 1990. Litterfall, leaf decomposition and litter colonization of *Tessaria integrifolia* (Compositae) in the Parana river floodplain. *Hydrobiologia* **203**: 45–52.

Neill, C. and Deegan, L.A. 1986. The effect of Mississippi River delta lobe development on the habitat composition and diversity of Louisiana coastal wetlands. *American Midland Naturalist* **116**: 296–303.

Nelson, C.W. and Nelson, J.R. 1984. The Central Valley riparian mapping project. In R.E. Warner and K.M. Hendrix (eds.) *California Riparian Systems*, pp. 307–13. University of California Press, Berkeley.

Nessel, J. 1978. Distribution and Dynamics of Organic Matter and Phosphorous in a Sewage Enriched Cypress Stand. M.S. thesis, University of Florida, Gainesville.

Newbold, J.D., Elwood, J.W., O'Neill, R.V. and Van Winkle, W. 1981. Measuring nutrient spiraling in streams. *Canadian Journal of Fisheries and Aquatic Sciences* **38**: 860–3.

Newsome, R.D., Kozlowski, T.T. and Tang, Z.C. 1982. Responses of *Ulmus americana* seedlings to flooding of soil. *Canadian Journal of Botany* **60**: 1688–95.

Niiyama, K, 1987. Distribution of salicaceous species and soil texture of habitats along the Ishikari River. *Japanese Journal of Ecology* **37**: 163–74.

Niiyama, K, 1989. Distribution of *Chosenia arbutifolia* and soil texture of habitats along the Satsunai River. *Japanese Journal of Ecology* **39**: 173–82.

Niiyama, K. 1990. The role of seed dispersal and seedling traits in colonization and coexistence of *Salix* species in a seasonally flooded habitat. *Ecological Research* **5**: 317–31.

Nilsson, C. 1986. Change in riparian plant community composition along two rivers in northern Sweden. *Canadian Journal of Botany* **64**: 589–92.

Nilsson, C. 1987. Distribution of stream-edge vegetation along a gradient of current velocity. *Journal of Ecology* **75**: 513–22.

Nilsson, C. 1991. Conservation management of riparian communities. In L. Hansson (ed.) *Ecological Principles of Nature Conservation*, pp. 352–72. Elsevier Science, Amsterdam.

Nilsson, C. and Grelsson, G. 1990. The effects of litter displacement on riverbank vegetation. *Canadian Journal of Botany* **68**: 735–41.

Nilsson, C. and Wilson, S.D. 1991. Convergence in plant community structure along disparate gradients: are lakeshores inverted mountainsides? *American Naturalist* **137**: 774–90.

Nilsson, C., Ekblad, A., Gardfjell, M. and Carlberg, B. 1991*a*. Long-term effects of river regulation on river margin vegetation. *Journal of Applied Ecology* **28**: 963–87.

Nilsson, C., Gardfjell, M. and Grelsson, G. 1991*b*. Importance of hydrochory in structuring plant communities along rivers. *Canadian Journal of Botany* **69**: 2631–3.

Nilsson, C., Grelsson, G., Dynesius, M., Johansson, M.E. and Sperens, U. 1991*c*. Small rivers behave like large rivers: effects of postglacial history on plant species richness along riverbanks. *Journal of Biogeography* **18**: 533–41.

Nilsson, C., Grelsson, G., Johansson, M. and Sperens, U. 1988. Can rarity and diversity be predicted in vegetation along river banks? *Biological Conservation* **44**: 201–12.

Nilsson, C., Grelsson, G., Johansson, M. and Sperens, U. 1989. Patterns of plant species richness along riverbanks. *Ecology* **70**: 77–84.

Nixon, E.S., Willett, R.L. and Cox, P.W. 1977. Woody vegetation of a virgin forest in an eastern Texas river bottom. *Castanea* **42**: 227–36.

Noble, C.A. and Palmquist, R.C. 1968. Meander growth in artificially straightened streams. *Iowa Academy of Science Proceedings* **75**: 234–42.

Noble, I.R. and Slatyer, R.O. 1980. The use of vital attributes to predict successional changes in plant communities subject to recurrent disturbances. *Vegetatio* **43**: 5–21.

Noble, M.G. 1979. The origin of *Populus deltoides* and *Salix interior* zones on point bars along the Minnesota River. *American Midland Naturalist* **102**: 59–67.

Nortcliff, S. and Thornes, J.B. 1988. The dynamics of a tropical floodplain environment with reference to forest ecology. *Journal of Biogeography* **15**: 49–59.

North, M.E.A. and Teversham, J.M. 1984. The vegetation of the floodplains of the Lower Fraser, Serpentine and Nicomekl Rivers, 1859–1890. *Syesis* **17**: 47–66.

Novitzki, R.P. 1978. Hydrologic characteristics of Wisconsin's wetlands and their influence on floods, stream flow, and sediment. In *Wetland Functions and Values: The State of Our Understanding*, pp. 377–88. American Water Resources Association.

Nummi, P. 1989. Simulated effects of the beaver on vegetation, invertebrates and ducks. *Annales Zoologici Fennici* **26**: 43–52.

Odgaard, A.J. 1987. Streambank erosion along two rivers in Iowa. *Water Resources*

Research **23**: 1225–36.

Ogawa, H. and Male, J.W. 1986. Simulating the flood mitigation role of wetlands. *Journal of Water Resources Planning and Management* **112**: 114–28.

Okagbue, C.O. and Abam, T.K.S. 1986. An analysis of stratigraphic control on river bank failure. *Engineering Geology* **22**: 231–45.

Oke, T.R. 1987. *Boundary Layer Climates*. London: Methuen.

Olson, T.E. and Knopf, F.L. 1986*a*. Naturalization of Russian-olive in the western United States. *Western Journal of Applied Forestry* **1**(3): 65–9.

Olson, T.E. and Knopf, F.L. 1986*b*. Agency subsidization of a rapidly spreading exotic. *Wildlife Society Bulletin* **14**: 492–3.

Olson, T.E. and Knopf, F.L. 1988. Patterns of relative diversity within riparian small mammal communities, Platte River watershed, Colorado. In R.C. Szaro, K.E. Severson and D.R. Patton (eds.) *Management of Amphibians, Reptiles, and Small Mammals in North America*. US Forest Service General Technical Report RM-166, pp. 379–86.

Omernik, J.M. 1987. Ecoregions of the coterminous United States. *Annals of the Association of American Geographers* **77**: 118–25.

Omernik, J.M., Abernathy, A.R. and Male, L.M. 1981. Stream nutrient levels and proximity of agricultural and forest land to streams: some relationships. *Journal of Soil and Water Conservation* **36**: 227–31.

Onstad, D.W. 1988. Population-dynamics theory: the roles of analytical, simulation, and supercomputer models. *Ecological Modelling* **43**: 111–24.

Orme, A.R. 1980. The need for physical geography. *Professional Geographer* **32**: 141–8.

Osborne, L.L. and Wiley, M.J. 1988. Empirical relationships between land use/cover and stream water quality in an agricultural watershed. *Journal of Environmental Management* **26**: 9–27.

Osterkamp, W.R. and Hupp, C.R. 1984. Geomorphic and vegetative characteristics along three northern Virginia streams. *Bulletin of the Geological Society of America* **95**: 1093–101.

Oulman, C. and Lohnes, R.A. 1985. Assessment of the relative contribution of channel versus sheet erosion in a Midwest river system. ISWRRI-144. Iowa State Water Resources Research Institute, Ames.

Paijmans, K. 1990. Wooded swamps in New Guinea. In A.E. Lugo, M. Brinson and S. Brown (eds.) *Forested Wetlands. Ecosystems of the World* 15, pp. 335–55. Elsevier, Amsterdam.

Palat, M. 1991. Model of matter flow in a representative ecosystem of the floodplain forest. In M. Penka, M. Vyskot, E. Klimo and F. Vasicek (eds.) *Floodplain Forest Ecosystem. II. After Water Management Measures*, pp. 265–78. Elsevier, Amsterdam.

Pallardy, S.G. and Kozlowski, T.T. 1981. Water relations of *Populus* clones. *Ecology* **62**: 159–69.

Parikh, A.K. 1989. Factors Affecting the Distribution of Riparian Tree Species in Southern California Chaparral Watersheds. Ph.D. dissertation, University of California, Santa Barbara.

Parker, J. 1950. The effects of flooding on the transpiration and survival of some southeastern forest tree species. *Plant Physiology* **25**: 453–560.

Parker, M., Wood, F.J., Smith, B.H. and Elder, R.G. 1985. Erosional downcutting in lower order riparian ecosystems: have historical changes been caused by removal of beaver. In R.R. Johnson, C.D. Ziebell, D.R. Patton, P.F. Ffolliott and R.H. Hamre (eds.) *Riparian Ecosystems and Their Management: Reconciling Conflicting Uses*. US Forest Service General Technical Report RM-120, pp. 35–8.

Parsons, D.A. 1963. Vegetative control of streambank erosion. In *Proceedings of the Federal Inter-Agency Sedimentation Conference*. US Department of Agriculture Miscellaneous Publication No. 970, pp. 130–6.

Pasche, E. and Rouve, G. 1985. Overbank flow with vegetatively roughened flood plains. *Journal of Hydraulic Engineering* **111**: 1262–78.

Pase, C.P. and Layser, E.F. 1977. Classification of riparian habitat in the Southwest. In R.R. Johnson and D.A. Jones (eds.) *Importance, Preservation and Management of Riparian Habitat: A Symposium*. US Forest Service General Technical Report RM-43, pp. 5–9.

Pastor, J., Bonde, J., Johnston, C.A. and Naiman, R.J. 1991. A mathematical treatment of the spatially dependent stabilities and beaver ponds. *Bulletin of the Ecological Society of America* **72**(2): 214.

Patten, D.T. 1984. Revegetation evaluation. In W.L. Graf *et al.* (eds.) *Issues Concerning Phreatophyte Clearing, Revegetation, and Water Savings Along the Gila River Arizona*. US Army Corps of Engineers, Los Angeles District Contract Report DACW09-83-M-2623, pp. 25–50.

Pautou, G. 1980. La dynamique de la végétation dans la vallée du Rhône entre Genève et Lyon. *La Végétation des Forêts Alluviales. Colloques Phytosociologiques* **9**: 81–91.

Pautou, G. 1983. Répercussions des aménagements hydroélectriques sur le dynamisme de la vegetation (l'example du Haut-Rhône français). *Revue de Géographie Alpine* **71**: 331–42.

Pautou, G. 1984. L'organisation des forêts alluviales dans l'axe Rhodanien entre Genève et Lyon; comparison avec des autres systèmes fluviaux. *Documents de Cartographie Ecologique* **27**: 43–64.

Pautou, G. 1988. Perturbations anthropiques et changements de végétation dans les systèmes fluviaux; l'organisation du paysage fluvial rhodanien entre Genève et Lyon. *Documents de Cartographie Ecologique* **31**: 73–96.

Pautou, G. and Bravard, J.P. 1982. L'incidence des activités humaines sur la dynamique de l'eau et l'evolution de la végétation dans la vallée du Haut-Rhône français. *Revue de Géographie de Lyon* **57**: 63–79.

Pautou, G. and Decamps, H. 1985. Ecological interactions between alluvial forests and hydrology of the Upper Rhone. *Archiv für Hydrobiologie* **104**: 13–37.

Pautou, G. and Girel, J. 1980. Les zones presentant un grand intérêt biologique dans la plaine alluviale du Rhône entre Genève et Lyon. *Les Forêts Alluviales. Colloques Phytosociologiques* **9**: 591–600.

Pautou, G. and Girel, J. 1982. Genese, évolution et disparition des pelouses calcaires dans la plaine alluviale du Rhône entre Genève et Lyon. *Les Pelouses Calcaires. Colloques Phytosociologiques* **11**: 239–42.

Pautou, G. and Girel, J. 1986. La végétation de la basse plaine de l'Ain: organisation spatiale et évolution. *Documents de Cartographie Ecologique* **29**: 75–96.

Pautou, G. and Girel, J. 1988. La phytosociologie, un outil performant pour l'étude des corridors fluviaux. *Phytosociologie et Paysages. Colloques Phytosociologiques* **17**: 415–23.

Pautou, G., Decamps, H., Amoros, C. and Bravard, J.P. 1985: Successions végétales dans les couloirs fluviaux: l'exemple de la plaine alluviale du Haut-Rhône français. *Bulletin d'Ecologie* **16**: 203–12.

Pautou, G., Girel. J. and Borel, J.L. 1992. Initial repercussions and hydroelectric developments in the French upper Rhone Valley: a lesson for predictive scenarios propositions. *Environmental Management* **16**: 231–42.

Pautou, G., Girel, J., Borel, J.L., Manneville, O. and Chalemont, J. 1991. Changes in flood-plain vegetation caused by damming: basis for a predictive diagnosis. In Ravera (ed.) *Terrestrial and Aquatic Ecosystems: Perturbation and Recovery*, pp. 126–34. Ellis Horward, Chichester.

Pautou, G., Girel, J., Lachet, B., and Ain, G. 1979. Récherches écologiques dans la vallée du Haut Rhône français. *Documents de Cartographie Ecologique* **22**: 5–63.

Payette, S. and Delwaide, A. 1991. Variations seculaires du niveau d'eau dans le bassin de la Rivière Boniface (Quebec Nordique): une analyse dendroecologique. *Géographie Physique et Quaternaire* **45**: 59–67.

Pearlstine, L., McKellar, H. and Kitchens, W. 1985. Modelling the impacts of a river diversion on bottomland forest communities in the Santee River floodplain, South Carolina. *Ecological Modelling* **29**: 283–302.

Peart, D.R. 1985. The quantitative representation of seed and pollen dispersal. *Ecology* **66**: 1081–3.

Penfound, W.T. 1947. An analysis of an elm-ash floodplain community near Norman, Oklahoma. *Proceedings of the Oklahoma Academy of Science* **27**: 59–60.

Pereira, J.S. and Kozlowski, T.T. 1977. Variations among woody angiosperms in response to flooding. *Physiologia Plantarum* **41**: 184–92.

Perkins, G. 1875. The vegetation of Illinois lowlands. *American Naturalist* **9**: 385–93.

Peterjohn, W.T. and Correll, D.L. 1984. Nutrient dynamics in an agricultural watershed: observations on the role of a riparian forest. *Ecology* **65**: 1466–75.

Peterjohn, W.T. and Correll, D.L. 1986. The effect of riparian forest on the volume and chemical composition of baseflow in an agricultural watershed. In Correll, D.L. (ed.) *Watershed Research Perspectives*, pp. 244–62. Smithsonian Institution Press, Washington.

Peterson, D.L. and Bazzaz, F.A. 1984. Photosynthetic and growth responses of silver maple (*Acer saccharinum* L.) seedlings to flooding. *American Midland Naturalist* **112**: 261–72.

Peterson, D.L. and Rolfe, G.L. 1982*a*. Nutrient dynamics and decomposition of litterfall in floodplain and upland forests of central Illinois. *Forest Science* **28**: 667–81.

Peterson, D.L. and Rolfe, G.L. 1982*b*. Seasonal variation in nutrients of floodplain and upland forest soils of central Illinois. *Journal of the Soil Science Society of America* **46**: 1310–15.

Peterson, D.L. and Rolfe, G.L. 1982*c*. Nutrient dynamics of herbaceous vegetation in upland and floodplain forest communities. *American Midland Naturalist* **107**: 325–39.

Petraitis, P.S., Latham, R.E. and Niesenbaum, R.A. 1989. The maintenance of species diversity by disturbance. *Quarterly Review of Biology* **64**: 393–418.

Petranka, J.W. and Holland, R. 1980. A quantitative analysis of bottomland communities in south-central Oklahoma. *Southwestern Naturalist* **25**: 207–14.

Pezeshki, S.R. and Chambers, J.L. 1986. Variation in flood-induced stomatal and photosynthetic responses of three bottomland tree species. *Forest Science* **32**: 914–23.

Phillips, J.D. 1988. Incorporating fluvial change in hydrologic simulations: a case study in coastal North Carolina. *Applied Geography* **8**: 25–36.

Phillips, J.D. 1989*a*. Fluvial sediment storage in wetlands. *Water Resources Bulletin* **25**: 867–73.

Phillips, J.D. 1989*b*. An evaluation of the factors determining the effectiveness of water quality buffer zones. *Journal of Hydrology* **107**: 133–45.

Phillips, J.D. 1989*c*. Nonpoint source pollution control effectiveness of riparian forests along a coastal plain river. *Journal of Hydrology* **110**: 221–37.

Phipps, R.L. 1972. Tree rings, stream runoff, and precipitation in central New York. US Geological Survey Professional Paper 800-B, pp. 259–64.

Phipps, R.L. 1979. Simulation of wetlands forest vegetation dynamics. *Ecological Modelling* **7**: 257–88.

Phipps, R.L. 1983. Streamflow of the Occoquan River in Virginia as reconstructed from tree-ring series. *Water Resources Bulletin* **19**: 735–43.

Pinay, G. 1988. Hydrobiological assessment of the Zambezi River system: a review. International Institute for Applied Systems Analysis Working Paper WP-88-089.

Pinay, G. and Decamps, H. 1988. The role of riparian woods in regulating nitrogen fluxes between the alluvial aquifer and surface water: a conceptual model. *Regulated Rivers* **2**: 507–16.

Pinay, G. and Labroue, L. 1986. Une station d'épuration naturelle des nitrates transportés par les nappes alluviales: l'aulnaie glutineuse. *Comptes Rendus*, Serie III **302**: 629–32.

Pinay, G., Decamps, H., Arles, C. and Lacassin-Seres, M. 1989. Topographic influence on carbon and nitrogen dynamics in riverine woods. *Archiv für Hydrobiologie* **114**: 401–14.

Pinay, G., Decamps, H., Chauvet, E. and Fustec, E. 1990. Functions of ecotones in fluvial systems. In R.J. Naiman and H. Decamps (eds.) *The Ecology and Management of Aquatic-Terrestrial Ecotones*. Man and the Biosphere Series, Vol. 4. UNESCO, Paris.

Pinay, G., Fabre, A., Vervier, P. and Gazelle, F. 1992. Control of C, N, P distribution in soils of riparian forests. *Landscape Ecology*, in press.

Pinay, G., Salewicz, K.A. and Kovacs, G. 1988. An attempt to facilitate water management issues in the Zambezi River basin using decision support systems. *Regulated Rivers: Research and Management* **2**: 559–63.

Platts, W.S. and Nelson, R.L. 1989. Stream canopy and its relationship to salmonid biomass in the intermountain West. *North American Journal of Fisheries Management* **9**: 446–57.

Poinsart, D., Bravard, J.P. and Caclin, M.C. 1989. Profil en long et granulometrie du lit des curs d'eau amenages: l'example du canal de Miribel (Haut-Rhône). *Revue de Géographie de Lyon* **64**: 240–51.

Ponce, V.M. 1987. Diffusion wave modeling of catchment dynamics. *Journal of Hydraulic Engineering* **112**: 716–27.

Popov, O.F. 1985. Certain principles of the regime and balance of ground water in forest and marsh landscapes of the right bank of the Pripyat River. *Soviet Meteorology and Hydrology* **11**: 70–7.

Popov, O.F. and Cherenkov, A.V. 1985. Total evaporation in pine and birch forests of the right bank of the Pripyat River. *Soviet Meteorology and Hydrology* **12**: 72–7.

Post, H.A. and de la Cruz, A.A. 1977. Litterfall, litter decomposition, and flux of particulate organic matter in a coastal plain stream. *Hydrobiologia* **55**: 201–7.

Prance, G.T. 1979. Notes on the vegetation of Amazonia III. The terminology of Amazonian forest types subject to inundation. *Brittonia* **31**: 26–38.

Pratt, T.R., Clewell, A.F. and Cleckley, W.O. 1989. Principal vegetation communities of the Choctawhatchee River flood plain, northwest Florida. In D.W. Fisk (ed.) *Wetlands: Concerns and Successes*, pp. 91–9. American Water Resources Association, Bethesda, Maryland.

Prax, A. 1991*a*. The hydrophysical properties of the soil and changes in them. In M. Penka, M. Vyskot, E. Klimo, and F. Vasicek (eds.) *Floodplain Forest Ecosystem*. II. *After Water Management Measures*, pp. 145–68. Elsevier, Amsterdam.

Prax, A. 1991*b*. Soil moisture content in connection with topography. In M. Penka, M. Vyskot, E. Klimo, and F. Vasicek (eds.) *Floodplain Forest Ecosystem*. II. *After Water Management Measures*, pp. 335–54. Elsevier, Amsterdam.

du Preez, P.J. and Venter, H.J.T. 1990. The phytosociology of the woody vegetation in the southern part of the Vredefort Dome area. Part I: communities of the plains, riverbanks and islands. *South African Journal of Botany* **56**: 631–6.

Rajaratnam, N. and Ahmadi, R.M. 1981. Hydraulics of channels with flood plains. *Journal of Hydraulics Research* **19**: 43–60.

Ranney, J.W. 1977. Forest island edges – their structure, development, and importance to regional forest ecosystem dynamics. EDFB/IBP-77/1. Oak Ridge National Laboratory, Tennessee.

Ranney, J.W. and Johnson, W.C. 1977. Propagule dispersal among forest islands in southeastern South Dakota. *Prairie Naturalist* **9**: 17–24.

Ranney, J.W., Bruner, M.C. and Levenson, J.B. 1981. The importance of edge in the structure and dynamics of forest islands. In R.L. Burgess and D.M. Sharpe (eds.) *Forest Island Dynamics in Man-Dominated Landscapes*, pp. 67–96. Springer-Verlag, New York.

Rasanen, M.E., Salo, J.S. and Kalliola, R. 1987. Fluvial perturbance in the western Amazon basin: regulation by long-term sub-Andean tectonics. *Science* **238**: 1398–401.

Rattray, J.M. 1961. Vegetation types of southern Rhodesia. *Kirkia* **2**: 68–93.

Reily, P.W. and Johnson, W.C. 1982. The effects of altered hydrologic regime on tree growth along the Missouri River in North Dakota. *Canadian Journal of Botany* **60**: 2410–23.

Reisner, M. 1986. *Cadillac Desert*. Viking Press, New York.

Rejmanek, M. 1977. The concept of structure in phytosociology with references to classification of plant communities. *Vegetatio* **35**: 55–61.

Rejmanek, M., Sasser, C.E. and Gosselink, J.G. 1987. Modeling of vegetation dynamics in the Mississippi river deltaic plain. *Vegetatio* **69**: 133–40.

Rex, K.D. and Malanson, G.P. 1990. The fractal shape of riparian forest patches. *Landscape Ecology* **4**: 249–58.

Rhoads, B.L. and Miller, M.V. 1990. Impact of riverine wetlands construction and operation on stream channel stability: conceptual framework for geomorphic assessment. *Environmental Management* **14**: 799–807.

Rhodes, J., Skau, C.M., Greenlee, D. and Brown, D.L. 1985. Quantification of nitrate uptake by riparian forests and wetlands in an undisturbed headwaters watershed. In R.R. Johnson, C.D. Ziebell, D.R. Patton, P.F. Ffolliott and R.H. Hamre (eds.) *Riparian Ecosystems and Their Management: Reconciling Conflicting Uses*. US Forest Service General Technical Report RM-120, pp. 175–9.

Rice, E.L. 1965. Bottomland forests of north-central Oklahoma. *Ecology* **46**: 708–14.

Richardot-Goulet, M., Castella, E. and Castella, C. 1987. Clarification and succession of former channels of the French upper Rhone alluvial plain using mollusca. *Regulated Rivers* **1**: 111–27.

Richoux, P and Castella, E. 1986. The aquatic Coleoptera of former riverbeds submitted to large hydrological fluctuations. In *Proceedings of the 2nd European Congress of Entomology*, pp. 129–32. Amsterdam.

Ridgway, R. 1872. Notes on the vegetation of the lower Wabash Valley. *American Naturalist* **6**: 658–65; 724–32.

Riegel, G.M., Svejcar, T.J., Blank, R.R. and Trent, J.D. 1991. Water and nutrient dynamics in a riparian montane meadow (abstract). *Bulletin of the Ecological Society of America* **72**: 230.

Rieley, J. and Page, S. 1990. *Ecology of Plant Communities*. Longman, Harlow, Essex.

Rikard, M.W. 1988. Hydrologic and Vegetative Relationships of the Congaree Swamp National Monument (South Carolina). Ph.D. dissertation, Clemson University, Clemson, South Carolina.

Risser, P.G. 1990. The ecological importance of land-water ecotones. In R.J. Naiman and H. Decamps, eds. *The Ecology and Management of Aquatic-Terrestrial Ecotones*, pp. 7–21. Man and the Biosphere Series, Vol. 4. UNESCO, Paris.

Risser, P.G., Karr, J.R. and Forman, R.T.T. 1984. *Landscape Ecology. Directions and Approaches*. Illinois Natural History Survey Special Publication No. 2, Champaign.

Roberts, J. and Ludwig, J.A. 1991. Riparian vegetation along current-exposure gradients in floodplain wetlands of the River Murray, Australia. *Journal of Ecology* **79**: 117–27.

Roberts, R.S. and Lant, C.L. 1988. Evaluating the environmental services of riparian wetlands as public goods: a program for agricultural land use in Iowa. Project 25706 Final Report, Iowa State Water Resources Research Institute, Ames.

Robertson, P.A., Weaver, G.T. and Cavanaugh, J.A. 1978. Vegetation and tree species patterns near the northern terminus of the southern floodplain forest. *Ecological Monographs* **48**: 249–67.

Robichaux, R. 1977. Geologic history of the riparian forests of California. In A. Sands (ed.) *Riparian Forests in California*, pp. 21–34. Institute of Ecology, Davis, California.

Robinson, T.W. 1958. Phreatophytes. US Geological Survey Water Supply Paper 1423.

Robinson, T.W. 1965. Introduction, spread and areal extent of saltcedar (*Tamarix*) in the western states. US Geological Survey Professional Paper 491-A: 1–12.

Robison, E.G. and Beschta, R.L. 1990*a*. Coarse woody debris and channel morphology interactions for undisturbed streams in southeast Alaska, USA. *Earth Surface Processes and Landforms* **15**: 149–56.

Robison, E.G. and Beschta, R.L. 1990*b*. Identifying trees in riparian areas that can provide coarse woody debris to streams. *Forest Science* **36**: 790–801.

Rohdenburg, H., Diekkruger, B. and Bork, H.R. 1986. Deterministic hydrological site and catchment models for the analysis of agroecosystems. *Catena* **13**: 119–37.

Rohm, C.M., Giese, J.W. and Bennett, C.C. 1987. Evaluation of an aquatic ecoregion classification of streams in Arkansas. *Journal of Freshwater Ecology* **4**: 127–40.

Rood, S.B. and Heinz-Milne, S. 1989. Abrupt downstream forest decline following river damming in southern Alberta. *Canadian Journal of Botany* **67**: 1744–9.

Rood, S.B. and Mahoney, J.M. 1990. Collapse of riparian forests downstream from dams in western prairies: probable causes and prospects for mitigation. *Environmental Management* **14**: 451–64.

Ross, H.H. 1963. Stream communities and terrestrial biomes. *Archiv für Hydrobiologie* **59**: 235–42.

Rostan, J.C., Amoros, C. and Juget, J. 1987. The organic content of the surficial sediment: a method for the study of ecosystems development in abandoned river channels. *Hydrobiologia* **148**: 45–62.

Rosenzweig, M.L. and Abramsky, Z. 1986. Centrifugal community organization. *Oikos* **46**: 339–48.

Roux, A.L. 1986. La gestion de l'eau et des milieux associés dans les plaines alluviales. Le cas de la riviere Ain, affluent du Rhône à l'amont de Lyon. *Documents de Cartographie Ecologique* **29**: 13–15.

Roux, A.L., Bravard, J.P., Amoros, C. and Pautou, G. 1989. Ecological changes of the French upper Rhone River since 1750. In G.E. Petts (ed.) *Historical Change of Large Alluvial Rivers: Western Europe*, pp. 323–50. Wiley, New York.

Ruddy, B.C. 1989. Use of a hydraulic potentiomanometer to determine groundwater gradients in a wetland, Colorado. In D.W. Fisk (ed.) *Wetlands: concerns and successes*, pp. 175–83. American Water Resources Association, Bethesda, Maryland.

Running, S.W. and Coughlan, J.C. 1988. A general model of forest ecosystem processes for regional applications, I. Hydrologic balance, canopy gas exchange and primary production processes. *Ecological Modelling* **42**: 125–54.

Rushton, B.T. 1988. Wetland Reclamation by Accelerating Succession. Ph.D. dissertation, University of Florida, Gainesville.

Salisbury, N.E., Knox, J.C. and Stephenson, R.A. 1968. *The Valleys of Iowa*. Department of Geography, University of Iowa, Iowa City.

Salo, J. 1990. External processes influencing origin and maintenance of inland water-land ecotones. In R.J. Naiman and H. Decamps (eds.) *The Ecology and Management of Aquatic-Terrestrial Ecotones*, pp. 37–64. Man and the Biosphere Series, Vol. 4. UNESCO, Paris.

Salo, J., Kalliola, R., Hakkinen, I., Makinen, Y., Niemela, P., Puhakka, M. and Coley, P.D. 1986. River dynamics and the diversity of Amazon lowland forest. *Nature* **322**: 254–8.

Sampson, H.C. 1930. Succession in the swamp forest formation in northern Ohio. *Ohio Journal of Science* **30**: 340–57.

Sanchez-Perez, J.M., Tremolieres, M. and Carbiener, R. 1991. Une station d'épuration naturelle des phosphates et nitrates apportés par les eaux de débordement du Rhin: la forêt alluviale à frêne et orme. *Comptes Rendus*, Serie III **312**: 395–402.

Sands, A. (ed.) 1977. *Riparian Forests in California*. Institute of Ecology Publication No. 15, Davis, California.

Sauer, J.D. 1988. *Plant Migration*. University of California Press, Berkeley.

Schlosser, I.J. and Karr, J.R. 1981*a*. Riparian vegetation and channel morphology impact on spatial patterns of water quality in agricultural watersheds. *Environmental Management* **5**: 233–43.

Schlosser, I.J. and Karr, J.R. 1981*b*. Water quality in agricultural watersheds: impact of riparian vegetation during baseflow. *Water Resources Bulletin* **17**: 233–40.

Schmithusen, J. 1963. Der Wissenschaftliche Landschaftsbegriff. *Mitteilungen der floristischsoziologischen Arbeitsgemeinschaft* NF-**10**: 9–19.

Schneider, R.L. and Sharitz, R.R. 1986. Seed bank dynamics in a southeastern riverine swamp. *American Journal of Botany* **73**: 1022–30.

Schneider, R.L. and Sharitz, R.R. 1988. Hydrochory and regeneration in a bald cypress – water tupelo swamp forest. Ecology **69**: 1055–63.

Schnitzler, A., Carbiener, R. and Sanchez-Perez, J.M. 1991. Variation in vernal species composition in alluvial forests of the Rhine valley, eastern France. *Journal of Vegetation Science* **2**: 485–90.

Schreiber, K-F. 1990. The history of landscape ecology in Europe. In I.S. Zonneveld and R.T.T. Forman (eds.) *Changing Landscapes: An Ecological Perspective*, pp. 21–33. Springer-Verlag, New York.

Schumm, S.A. and Meyer, D.F. 1979. Morphology of alluvial rivers of the Great Plains. *Great Plains Agricultural Council* **91**: 9–14.

Sebenik, P.G. and Thames, J.L. 1967. Water consumption by phreatophytes. *Progressive Agriculture in Arizona* **19**: 10–11.

Sedell, J.R. and Froggatt, J.L. 1984. Importance of streamside forests to large rivers: the isolation of the Willamette River, Oregon, U.S.A., from its floodplain by snagging and streamside forest removal. *Verhandlungen Internationale Vereiningung für Theoretische und Angewandte Limnologie* **22**: 1828–34.

Sedgwick, J.A. and Knopf, F.L. 1986. Cavity-nesting birds and the cavity-tree resource in Plains cottonwood bottomlands. *Journal of Wildlife Management* **50**: 247–52.

Sedgwick, J.A. and Knopf, F.L. 1987. Breeding bird response to cattle grazing of a cottonwood bottomland. *Journal of Wildlife Management* **51**: 230–7.

Sedgwick, J.A. and Knopf, F.L. 1989. Demography, regeneration, and future projections for a bottomland cottonwood community. In R.R. Sharitz and J.W. Gibbons (eds.) *Freshwater Wetlands and Wildlife*. USA Department of Energy Symposium Series No. 61, CONF-8603101, Oak Ridge, pp. 249–66.

Sedgwick, J.A. and Knopf, F.L. 1990. Habitat relationships and nest site characteristics of cavity-nesting birds in cottonwood floodplains. *Journal of Wildlife Management* **54**: 112–24.

Sedgwick, J.A. and Knopf, F.L. 1991. Prescribed grazing as a secondary impact in a western riparian floodplain. *Journal of Range Management* **44**: 369–73.

Sena Gomes, A.R. and Kozlowski, T.T. 1980*a*: Growth responses and adaptations of

Fraxinus pennsylvanicus to flooding. *Plant Physiology* **66**: 267–71.

Sena Gomes, A.R. and Kozlowski, T.T. 1980*b*. Responses of *Melaleuca quinquenervia* seedlings to flooding. *Physiologia Plantarum* **49**: 373–7.

Sena Gomes, A.R. and Kozlowski, T.T. 1980*c*. Effects of flooding on *Eucalyptus camaldulensis* and *Eucalyptus globulus* seedlings. *Oecologia* **46**: 139–42.

Sengupta, S., Lee, S.S., and Fu, L.-Q. 1986. Time dependent three dimensional simulation of flows in shallow domains with vegetative obstruction. *Applied Mathematical Modelling* **10**: 2–10.

Shankman, D. and Drake, L.G. 1990. Channel migration and regeneration of bald cypress in western Tennessee. *Physical Geography* **11**: 343–52.

Shanks, B.D. 1974. The American Indian and Missouri River water development. *Water Resources Bulletin* **10**: 573–9.

Sharitz, R.R. and Lee, L.C. 1985. Limits on regeneration processes in southeastern riverine wetlands. In R.R. Johnson, C.D. Ziebell, D.R. Patton, P.F. Ffolliott and R.H. Hamre (eds.) *Riparian Ecosystems and Their Management: Reconciling Conflicting Uses*. US Forest Service General Technical Report RM-120, pp. 139–43.

Shaw, R.K. 1976. A taxonomic and ecologic study of the riverbottom forest on St. Mary River, Lee Creek, and Belly River in southwestern Alberta, Canada. *Great Basin Naturalist* **36**: 243–71.

Shelford, V.E. 1954. Some lower Mississippi Valley flood plain biotic communities; their age and elevation. *Ecology* **35**: 126–42.

Shelyag-Sosonko, Yu.R., Sipaylova, L.M., Solomakha, V.A. and Mirkin, B.M. 1987. Meadow vegetation of the Desna River floodplain (Ucraine, USSR). *Folia Geobotanika and Phytotaxonomica* **22**: 113–69.

Sheppe, W.A. 1972. The annual cycle of small-mammal populations on a Zambian floodplain. *Journal of Mammalogy* **53**: 445–60.

Sheppe, W.A. 1985. Effects of human activities on Zambia's Kafue flats ecosystem. *Environmental Conservation* **12**: 49–57.

Sheppe, W.A. and Osborne, T. 1971. Patterns of use of a flood plain by Zambian mammals. *Ecological Monographs* **41**: 179–205.

Shields, F.D. 1991. Woody vegetation and riprap stability along the Sacramento River mile 84.5-119. *Water Resources Bulletin* **27**: 527–36.

Shirohammadi, A., Sheridan, J.M. and Rasmussen, L.E. 1986. Hydrology of alluvial stream channels in southern coastal plain watersheds. *Transactions of the ASAE* **29**: 135–42.

Shroder, J.F. Jr. and Butler, D.R. 1987. Tree-ring analysis in the earth sciences. In G.C. Jacoby and J.W. Hornbeck (eds.) *Proceedings of the International Symposium on Ecological Aspects of Tree-Ring Analysis*. US Department of Energy CONF-8608144, pp. 186–212.

Shugart, H.H. and West, D.C. 1977. Development of an Appalachian deciduous forest succession model and its application to assessment of the impact of chestnut blight. *Journal of Environmental Management* **5**: 161–79.

Shull, C.A. 1922. The formation of a new island in the Mississippi River. Ecology **3**: 202–6.

Shull, C.A. 1944. Observations of general vegetation changes on a river island in the Mississippi River. *American Midland Naturalist* **32**: 771–6.

Shunk, I.V. 1939. Oxygen requirements for germination of seeds of *Nyssa aquatica* – tupelo gum. *Science* **90**: 565–6.

Shure, D.J. and Gottschalk, M.R. 1985. Litter-fall patterns within a floodplain forest. *American Midland Naturalist* **114**: 98–111.

Shure, D.J., Gottschalk, M.R. and Parsons, K.A. 1986. Litter decomposition processes in a floodplain forest. *American Midland Naturalist* **115**: 314–27.

Siegrist, R. 1913. *Die Auenwalder der Aare*. Aarau.

Sigafoos, R.S. 1964. Botanical evidence of floods and floodplain deposition. US Geological Survey Professional Paper 485-A.

Sigafoos, R.S. and Sigafoos, M.D. 1966. Flood history told by tree growth. *Natural History* **75**: 50–5.

Simberloff, D. and Abele, L.G. 1976. Island biogeography theory and conservation practice. *Science* **191**: 285–6.

Simon, A. 1989. The discharge of sediment in channelized alluvial streams. *Water Resources Bulletin* **25**: 1177–88.

Simon, A. and Hupp, C.R. 1986. Channel widening characteristics and bank slope development along a reach of Cane Creek, West Tennessee. US Geological Survey Water Supply Paper 2290, pp. 113–26.

Simon, A. and Hupp, C.R. 1987. Geomorphic and vegetative recovery processes along modified Tennessee streams: an interdisciplinary approach to disturbed fluvial systems. In *Forest Hydrology and Watershed Management*, pp. 251–62. IAHS Publication No. 167. International Association of Hydrological Sciences, Washington.

Simpson, C.D. 1975. A detailed vegetation study on the Chobe River in N.E. Botswana. *Kirkia* **10**: 185–227.

Sinitsyn, M.G. and Rusanov, A.V. 1990. European beaver impact on small rivers' valley and channel relief in the Vetluga-Unzha woodlands. *Geomorfologiya* **1990**(1): 85–91. (In Russian, summary in English.)

Sioli, H. 1984. *The Amazon*. W. Junk, Dordrecht.

Sipp, S.K. and Bell, D.T. 1974. The response of net photosynthesis to flood conditions in seedlings of *Acer saccharinum* (silver maple). University of Illinois Forestry Research Report 74–9.

Sklar, F.H. and Costanza, R. 1991. The development of dynamic spatial models for landscape ecology: a review and prognosis. In M.G. Turner and R.H. Gardner (eds.) *Quantitative Methods in Landscape Ecology*, pp. 239–88. Springer-Verlag, New York.

Skoglund, S.J. 1990. Seed dispersing agents in two regularly flooded river sites. *Canadian Journal of Botany* **68**: 754–60.

Skoglund, J. and Verwijst, T. 1989. Age structure of woody species populations in relation to seed rain, germination and establishment along the river Dalaven, Sweden. *Vegetatio* **82**: 25–34.

Slater, F.M., Curry, P., and Chadwell, C. 1987: A practical approach to the evaluation of the conservation status of vegetation in river corridors in Wales. *Biological Conservation* **40**: 53–68.

Smith, D.G. 1976. Effect of vegetation on lateral migration of a glacial meltwater river. *Bulletin of the Geological Society of America* **87**: 857–60.

Smith, R.H. and Shields, F.D. 1990. Effects of clearing and snagging on physical conditions of rivers. In *Proceedings, Twentieth Mississippi Water Resources Conference*, pp. 41–51. Mississippi State Water Resources Research Institute, Jackson.

Smith, R.J., Hancock, N.H. and Ruffini, J.L. 1990. Flood flow through tall vegetation. *Agricultural Water Management* **18**: 317–32.

Smith, S.D., Nachlinger, J.L., Wellington, A.B. and Fox, C.A. 1989. Water relations of obligate riparian plants as a function of streamflow diversion on the Bishop Creek watershed. In D.L. Abell (ed.) *Proceedings of the California Riparian Systems Conference.* US Forest Service General Technical Report PSW-110, pp. 360–5.

Smith, S.D., Wellington, A.B., Nachlinger, J.L. and Fox, C.A. 1991. Functional responses of riparian vegetation to streamflow diversion in the eastern Sierra Nevada. *Ecological Applications* **1**: 89–97.

Sollers, S.C. 1974. Substrate conditions, community structure, and succession in a portion of the floodplain of Wissahickon Creek. *Bartonia* **42**: 24–42.

Sophocleous, M., Townsend, M.A., Vogler, L.D., McClain, T.J., Marks, E.T. and Coble, G.R. 1988. Experimental studies in stream-aquifer interaction along the Arkansas River in central Kansas – field testing and analysis. *Journal of Hydrology* **98**: 249–73.

Space, M.L. 1989. Hydrology of Bishop Creek, Inyo County, California: an Isotopic Analysis. MS thesis, University of Nevada, Las Vegas.

Sparks, R.E., Bayley, P.B., Kohler, S.L. and Osborne, L.L. 1990. Disturbance and recovery of large floodplain rivers. *Environmental Management* **14**: 699–709.

Springer, F.M., Ullrich, C.R. and Hagerty, D.J. 1985. Streambank stability. *Journal of Geotechnical Engineering* **111**: 624–40.

Star, J. and Estes, J. 1990. *Geographic Information Systems.* Prentice Hall, Englewood Cliffs, New Jersey.

Steinblums, I.J. and Leven, A.A. 1985. Riparian area management in the Pacific Southwest Region, USDA Forest Service. In R.R. Johnson, C.D. Ziebell, D.R. Patton, P.F. Ffolliott and R.H. Hamre (eds.) *Riparian Ecosystems and Their Management: Reconciling Conflicting Uses.* US Forest Service General Technical Report RM-120, pp. 507–9.

Sternberg, H.A. 1975. The Amazon River of Brazil. *Erdkunde Wiseen* **40**: (xii + 74 pp.).

Stevens, L.E. 1989. Mechanisms of Riparian Plant Community Organization and Succession in the Grand Canyon, Arizona. Ph.D. dissertation, Northern Arizona University, Flagstaff.

Stevens, L.E. and Waring, G.L. 1985. The effects of flooding on the riparian plant communities in Grand Canyon. In R.R. Johnson, C.D. Ziebell, D.R. Patton, P.F. Ffolliott, and R.H. Hamre (eds.) *Riparian Ecosystems and Their Management: Reconciling Conflicting Uses.* US Forest Service General Technical Report RM-120, pp. 81–6.

Stine, S., Gaines, E. and Vorster, P. 1984. Destruction of riparian systems due to water development in the Mono Lake watershed. In R.E. Warner and K.M. Hendrix (eds.) *California Riparian Systems*, pp. 58–67. University of California Press, Berkeley.

Stockton, C.W. 1975. Long term streamflow records reconstructed from tree rings. *Papers of the Laboratory for Tree-Ring Research* 5. University of Arizona Press, Tucson.

Stockton, C.W. and Fritts, H.C. 1971. Augmentation of hydrologic records using tree-ring data. In *Hydrology and water resources in Arizona and the Southwest*, pp. 1–12. Arizona Academy of Sciences, Tuscon.

Stockton, C.W. and Fritts, H.C. 1973. Long-term reconstruction of water level changes for Lake Athabasca by analysis of tree rings. *Water Resources Bulletin* **9**: 1006–27.

Stone, E.C., Cavallaro, J.I. and Stromberg, L.P. 1984. Spatial vegetation units used with a description method based on two levels of resolution to provide the requisite structural information for vegetation preservation. In R.E. Warner and K.M. Hendrix (eds.) *California Riparian Systems*, pp. 340–6. University of California Press, Berkeley.

Storm, G.L., Andrews, R.D., Phillips, R.L., Bishop, R.A., Sniff, D.B. and Tester, J.R. 1976. Morphology, reproduction, dispersal and mortality of midwestern red fox populations. *Wildlife Monographs* **49**: 5–82.

Streng, D.R., Glitzenstein, J.S. and Harcombe, P.A. 1989. Woody seedling dynamics in an East Texas floodplain. *Ecological Monographs* **59**: 177–204.

Stromberg, J.C. and Katibah, E.F. 1984. An application of the spatial-aggregation method to the description of riparian vegetation. In R.E. Warner and K.M. Hendrix (eds.) *California Riparian Systems*, pp. 347–55. University of California Press, Berkeley.

Stromberg, J.C. and Patten, D.T. 1989. Early recovery of an eastern Sierra Nevada riparian system after 40 years of stream diversion. In D.L. Abell (ed.) *Proceedings of the California Riparian Systems Conference*. US Forest Service General Technical Report PSW-110, pp. 399–404.

Stromberg, J.C. and Patten, D.T. 1990. Riparian vegetation instream flow requirements: a case study from a diverted stream in the eastern Sierra Nevada, California, USA. *Environmental Management* **14**: 185–94.

Stuber, P.R. and Sather, J.H. 1986. Research gaps in assessing wetland functions. *Transactions of the North American Wildlife and Natural Resources Conference* **49**: 304–11.

Sundeen, K.D., Leaf, C.F. and Bostrom, G.M. 1989. Hydrologic functions of subalpine wetlands in Colorado. In *Wetlands: Concerns and Successes*, pp. 401–13. American Water Resources Association, Bethesda, Maryland.

Svendson, G.E. 1980. Seasonal change in feeding patterns of beaver in southeastern Ohio. *Journal of Wildlife Management* **44**: 285–90.

Swanson, F.J., Gregory, S.V., Sedell, J.R. and Campbell, A.G. 1982. Land-water interactions: the riparian zone. In Edmonds, R.L. (ed.) *Analysis of Coniferous Forest Ecosystems in the Western United States*, pp. 267–91. Hutchinson Ross, Stroudsburg, Pennsylvania.

Swanson, F.J. and Lienkaemper, G.W. 1978. Physical consequences of large organic debris in Pacific Northwest streams. US Forest Service General Technical Report PNW-69.

Swanson, F.J., Lienkaemper, G.W. and Sedell, J.R. 1976. History, physical effects, and management implications of large organic debris in western Oregon streams. US Forest Service General Technical Report PNW-56.

Szaro, R.C. 1990. Southwestern riparian plant communities: site characteristics, tree species distributions, and size-class structures. *Forest Ecology and Management* **33/34**: 315–34.

Szaro, R.C. and Belfit, S.C. 1986. Herpetofaunal use of a desert riparian island and its adjacent scrub habitat. *Journal of Wildlife Management* **50**: 752–61.

Szaro, R.C. and DeBano, L.F. 1985. The effects of streamflow modification on the development of a riparian ecosystem. In R.R. Johnson, C.D. Ziebell, D.R. Patton, P.F. Ffolliott and R.H. Hamre (eds.) *Riparian Ecosystems and Their Management: Reconciling Conflicting Uses*. US Forest Service General Technical Report RM-120, pp. 211–15.

Szaro, R.C. and King, R.M. 1990. Sampling intensity and species richness: effects on delineating southwestern riparian plant communities. *Forest Ecology and Management* **33/34**: 335–9.

Tabacchi, E. and Planty-Tabacchi, A.-M. 1990. Évolution longitudinale de la végétation du corridor de l'Adour. *Botanica Pirenaico-Cantabrica* **1990**: 455–68.

Tabacchi, E., Planty-Tabacchi, A.-M. and Decamps, O. 1990. Continuity and discontinuity of the riparian vegetation along a fluvial corridor. *Landscape Ecology* **5**: 9–20.

Tang, Z.C. and Kozlowski, T.T. 1982. Some physiological and morphological responses of *Quercus macrocarpa* seedlings to flooding. *Canadian Journal of Forest Research* **12**: 196–202.

Tanner, J.T. 1986: Distribution of tree species in Louisiana bottomland forests. *Castanea* **51**: 168–74.

Tate, C.M. and Gurtz, M.E. 1986. Comparison of mass loss, nutrients, and invertebrates associated with elm leaf litter decomposition in perennial and intermittent reaches of tallgrass prairie streams. *Southwestern Naturalist* **31**: 511–20.

Taylor, J.R. 1985. Community Structure and Primary Productivity of Forested Wetlands in Western Kentucky. Ph.D. dissertation, University of Louisville.

Tchou, Y.-T. 1951. Études écologiques et phytosociologiques sur les forêts riveraines du Bas-Languedoc. *Vegetatio* **1**: 2–28; 93–128; 217–57; 347–83.

Telenius, A. and Torstensson, P. 1991. Seed wings in relation to seed size in the genus *Spergularia*. *Oikos* **61**: 216–22.

Teskey, R.O. and Hinckley, T.M. 1977*a*. Impact of water level changes on woody riparian and wetland communities. Vol. I: Plant and soil responses to flooding. US Fish and Wildlife Service OBS-77/58.

Teskey, R.O. and Hinckley, T.M. 1977*b*. Impact of water level changes on woody riparian and wetland communities. Vol. II: Southern forest region. US Fish and Wildlife Service OBS-77/59.

Teskey, R.O. and Hinckley, T.M. 1978*a*. Impact of water level changes on woody riparian and wetland communities. Vol. IV: Eastern deciduous forest region. US Fish and Wildlife Service OBS-78/87.

Teskey, R.O. and Hinckley, T.M. 1978*b*. Impact of water level changes on woody riparian and wetland communities. Vol. V: Northern forest region. US Fish and Wildlife Service OBS-78/88.

Teskey, R.O. and Hinckley, T.M. 1978*c*. Impact of water level changes on woody riparian and wetland communities. Vol. VI: Plains grassland region. US Fish and Wildlife Service OBS-78/89.

Tessier, C., Maire, A. and Aubin, A. 1981. Étude de la végétation des zones riveraines de l'archipel des Centiles du fleuve Saint Laurent, Quebec. *Canadian Journal of Botany* **59**: 1526–36.

Teversham, J.M. and Slaymaker, D. 1976. Vegetation composition in relation to flood frequency in Lilloet River Valley, British Columbia. *Catena* **3**: 191–201.

Thebaud, C. and Debussche, M. 1991. Rapid invasion of *Fraxinus ornus* L. along the

Herault River system in southern France: the importance of seed dispersal by water. *Journal of Biogeography* **18**: 7–12.
Thomas, D.L. and Beasley, D.B. 1986. A physically-based forest hydrology model I: Development and sensitivity of components. *Transactions of the American Society of Agricultural Engineers (ASAE)* **29**: 962–72.
Thompson, K. 1961. Riparian forests of the Sacramento Valley, California. *Annals of the Association of American Geographers* **51**: 294–315.
Thompson, K. 1977. Riparian forests of the Sacramento Valley, California. In A. Sands (ed.) *Riparian Forests in California*, pp. 35–8. Institute of Ecology, Davis, California.
Thornes, J.B. 1990. *Vegetation and Erosion*. John Wiley, Chichester.
Tilman, D. 1988. *Plant Strategies and the Dynamics and Structure of Plant Communities*. Princeton University Press, Princeton.
Titus, J.H. 1990. Microtopography and woody plant regeneration in a hardwood floodplain swamp in Florida. *Bulletin of the Torrey Botanical Club* **117**: 429–37.
Trent, R.E. and Brown, S.A. 1984. An overview of factors affecting river stability. *Transportation Research Record* **950**: 156–63.
Trimble, S.W. 1981. Changes in sediment storage in the Coon Creek Basin, Driftless Area, Wisconsin, 1853–1975. *Science* **214**: 181–3.
Trimble, S.W. 1983. A sediment budget for Coon Creek basin in the Driftless Area, Wisconsin, 1853–1977. *American Journal of Science* **283**: 454–74.
Trimble, S.W. and Lund, S.W. 1982. Soil conservation and the reduction of erosion and sedimentation in the Coon Creek Basin, Wisconsin. US Geological Survey Professional Paper 1234.
Trimble, S.W. and Knox, J.C. 1984. Comment on 'Erosion, redeposition, and delivery of sediment to midwestern streams' by D.C. Wilkin and S.J. Hebel. *Water Resources Research* **20**: 1317–8.
Triska, F.J. 1984. Role of woody debris in modifying channel geometry and riparian areas of a large lowland river under pristine conditions: a historical case study. *Verhandlungen Internationale Vereiningung für Theoretische und Angewandte Limnologie* **22**: 1876–92.
Troll, C. 1939. Luftbildplan und ökologische Bodenforschung. *Zeitschrift der Gesellschaft für Erdkunde, Berlin* **1939**: 241–98.
Troll, C. 1968. Landschaftsokologie. In R. Tuxen (ed.) *Pflanzensoziologie und Landschaftsökologie*, pp. 1–21. W. Junk, The Hague.
Trush, W.J., Connor, E.C. and Knight, A.W. 1989. Alder establishment and channel dynamics in a tributary of the South Fork Eel River, Mendocino County, California. In D.L. Abell (ed.) *Proceedings of the California Riparian Systems Conference*. US Forest Services General Technical Report PSW-110, pp. 14–21.
Tsukahara, H. and Kozlowski, T.T. 1985. Importance of adventitious roots to growth of flooded *Platanus occidentalis* seedlings. *Plant and Soil* **88**: 123–32.
Turner, L.M. 1936. Ecological studies in the lower Illinois River valley. *Botanical Gazette* **97**: 689–727.
Turner, M.G. (ed.) 1987. *Landscape Heterogeneity and Disturbance*. Springer Verlag, New York.
Turner, M.G. 1990. Landscape changes in nine rural counties in Georgia. *Photogrammetric Engineering and Remote Sensing* **56**: 379–86.
Turner, M.G., Costanza, R. and Sklar, F.H. 1989. Methods to evaluate the

performance of spatial simulation models. *Ecological Modelling* **48**: 1–18.

Turner, M.G. and Gardner, R.H. (eds.) 1991. *Quantitative Methods in Landscape Ecology*. Springer-Verlag, New York.

Turner, R.E., Forsythe, S.W. and Craig, N.J. 1981. Bottomland hardwood forest land resources of the southeastern United States. In J.R. Clark and J. Benforado (eds.) *Wetland of Bottomland Hardwood Forests*, pp. 13–28. Elsevier, Amsterdam.

Turner, R.M. 1974. Quantitative and historical evidence of vegetation changes along the upper Gila River, Arizona. US Geological Survey Professional Paper 655-H.

Turner, R.M. and Karpiscak, M.M. 1980. Recent vegetation changes along the Colorado River between Glen Canyon Dam and Lake Mead, Arizona. US Geological Survey Professional Paper 1132.

Turner, S.J., O'Neill, R.V., Conley, W., Conley, M.R. and Humphries, H.C. 1991. Pattern and scale: statistics for landscape ecology. In M.G. Turner and R.H. Gardner (eds.) *Quantitative Methods in Landscape Ecology*, pp. 17–49. Springer-Verlag, New York.

Uhrecky, I., Smolik, Z., Havlicek, V. and Mrvka, R. 1985. Radiation, temperature and rainfall regimes of the floodplain forest ecosystem. In M. Penka, M. Vyskot, E. Klimo and F. Vasicek (eds.) *Floodplain Forest Ecosystem*. I. *Before Water Management Measures*, pp. 33–59. Elsevier, Amsterdam.

Ulmer, R.L. 1990. Tennessee-Tombigbee Waterway canal section effect on adjacent bottomland hardwoods. In *Proceedings, Twentieth Mississippi Water Resources Conference*, pp. 92–7. Mississippi State Water Resources Research Institute, Jackson.

US Army Corps of Engineers. 1991. Mississippi reforestation project to yield information about hydrology and vegetation. *Wetlands Research Program Bulletin* **1**(2):4–5.

Vale, T.R. and Parker, A.J. 1980. Biogeography: research opportunities for geographers. *Professional Geographer* **32**: 149–57.

VanAuken, O.W. and Bush, J.K. 1985. Secondary succession on terraces of the San Antonio River. *Bulletin of the Torrey Botanical Club* **112**: 158–66.

van der Pijl, L. 1982. *Principles of Dispersal in Higher Plants*. Springer-Verlag, Berlin.

van der Valk, A.G. 1981. Succession in wetlands – a Gleasonian approach. *Ecology* **62**: 688–96.

Vanek, V. 1991. Riparian zone as a source of phosphorous for a groundwater-dominated lake. *Water Research* **25**: 409–18.

van Hylckama, T.E.A. 1974. Water use by saltcedar as measured by the water budget method. US Geological Survey Professional Paper 491-E.

Vannote, R.L., Minshall, G.W., Cummins, K.W., Sedell, J.R. and Cushing, C.E. 1980. The river continuum concept. *Canadian Journal of Fisheries and Aquatic Sciences* **37**: 130–7.

Van Sickle, J. and Gregory, S.V. 1990. Modeling inputs of large woody debris to streams from falling trees. *Canadian Journal of Forest Research* **20**: 1593–601.

Vasicek, F. 1985. Natural conditions of floodplain forests. In M. Penka, M. Vyskot, E. Klimo and F. Vasicek (eds.) *Floodplain Forest Ecosystem*. I. *Before Water Management Measures*, pp. 13–29. Elsevier, Amsterdam.

Vasicek, F. 1991. Changes in the structure and biomass of the herb layer under the conditions of a medium moisture gradient. In M. Penka, M. Vyskot, E. Klimo and F. Vasicek (eds.) *Floodplain Forest Ecosystem.* II. *After Water Management Measures*, pp. 197–228. Elsevier, Amsterdam.

Vasicek, F. and Pivec, J. 1991*a*. The meteorological conditions in southern Moravia following the control of flooding in floodplain forests. In M. Penka, M. Vyskot, E. Klimo and F. Vasicek (eds.) *Floodplain Forest Ecosystem.* II. *After Water Management Measures*, pp. 75–80. Elsevier, Amsterdam.

Vasicek, F. and Pivec, J. 1991*b*. Microclimate in floodplain forests. In M. Penka, M. Vyskot, E. Klimo and F. Vasicek (eds.) *Floodplain Forest Ecosystem.* II. *After Water Management Measures*, pp. 97–102. Elsevier, Amsterdam.

Vickers, G.T. 1991. Spatial patterns and travelling waves in population genetics. *Journal of Theoretical Biology* **150**: 329–37.

Viereck, L.A. 1970: Forest succession and soil development adjacent to the Chena River in interior Alaska. *Arctic and Alpine Research* **2**: 1–26.

Viles, H.A. (ed.) 1988. *Biogeomorphology*. Blackwell, Oxford.

Vyskot, M. 1976*a*. Biomass production of the tree layer in a floodplain forest near Lednice. In H.E. Young (ed.) *Oslo Biomass Studies*, pp. 175–202. University of Maine, Orono.

Vyskot, M. 1976*b*. Floodplain forest in biomass. In H.E. Young (ed.) *Oslo Biomass Studies*, pp. 203–29. University of Maine, Orono.

Vyskot, M. 1985. The tree layer. In M. Penka, M. Vyskot, E. Klimo and F. Vasicek (eds.) *Floodplain Forest Ecosystem.* I. *Before Water Management Measures*, pp. 81–120. Elsevier, Amsterdam.

Wakeman, T.H. and Fong, C.C. 1984. Section 404 jurisdictional determinations in riparian systems. In R.E. Warner and K.M. Hendrix (eds.) *California Riparian Systems*, pp. 899–903. University of California Press, Berkeley.

Waldemarson Jensen, E. 1979. Successions in relationship to lagoon development in the Laitaure delta, North Sweden. *Acta Phytogeographica Suecica* **66**: 1–128.

Walker, K.F. 1985. A review of the ecological effects of river regulation in Australia. *Hydrobiologia* **125**, 111–29.

Walker, L.R. and Chapin, F.S. 1986. Physiological controls over seedling growth in primary succession on an Alaskan floodplain. *Ecology* **67**: 1508–23.

Walker, L.R., Zasada, J.C. and Chapin, F.S. 1986. The role of life history processes in primary succession on an Alaskan floodplain. *Ecology* **67**: 1243–53.

Walsh, S.J. 1985. GIS for natural resource management. *Journal of Soil and Water Conservation* **40**: 202–5.

Walsh, S.J., Lightfoot, D.R. and Butler, D.R. 1987. Recognition and assessment of error in geographic information systems. *Photogrammetric Engineering and Remote Sensing* **53**: 1423–30.

Walters, M.A., Teskey, R.O. and Hinckley, T.M. 1980. Impact of water level changes on woody riparian and wetland communities. Vol. VIII: Pacific Northwest and Rocky Mountain regions. US Fish and Wildlife Service OBS-78/94.

Ware, G.H. and Penfound, W.T. 1949. The vegetation of the lower levels of the floodplain of the South Canadian River in central Oklahoma. *Ecology* **30**: 478–84.

Warner, R.E. and Hendrix, K.M. eds. 1984. *California Riparian Systems*. University of California Press, Berkeley.

Warren, D.K. and Turner, R.M. 1975. Saltcedar (*Tamarix chinensis*) seed production, seedling establishment, and response to inundation. *Journal of the Arizona Academy of Science* **10**: 135–44.

Wassen, M.J., Barendregt, A., Palczynski, A., de Smidt, J.T. and de Mars, H. 1990. The relationship between fen vegetation gradients, groundwater flow and flooding in an undrained valley mire at Biebrza, Poland. *Journal of Ecology* **78**: 1106–22.

Weaver, J.E. 1960. Floodplain vegetation of the central Missouri valley and contacts of woodland with prairie. *Ecological Monographs* **30**: 37–64.

Weaver, J.E., Hanson, H.C. and Aikman, J.M. 1925. Transect method of studying woodland vegetation along streams. *Botanical Gazette* **80**: 168–87.

Webb, T.III, Cushing, E.J. and Wright, H.E. 1983. Holocene changes in the vegetation of the Midwest. In H.E. Wright (ed.) *Late-Quaternary Environments of the United States*, vol. 2 *The Holocene*. University of Minnesota Press, Minneapolis.

Weiss, D., Carbiener, R. and Tremolieres, M. 1991. Biodisponibilité comparée du phosphore en fonction des substrats et de la fréquence des inondations dans trois forêts alluviales rhenanes de la plaine d'Alsace. *Comptes Rendus*, serie III **313**: 245–51.

Welcomme, R.L. 1979. *Fisheries Ecology of Floodplain Rivers*. Longman, New York.

Wendland, W.M. and Watson-Stegner, D. 1983. A technique to reconstruct river discharge history from tree-rings. *Water Resources Bulletin* **19**: 175–81.

Wenger, E.L., Zinke, A. and Gutzweiler, K.-A. 1990. Present situation of the European floodplain forests. *Forest Ecology and Management* **33/34**: 5–12.

Westman, W.E. 1983. Xeric Mediterranean-type shrubland associations of Alta and Baja California and the community/continuum debate. *Vegetatio* **52**: 3–19.

Wheeler, R.H. and Kapp, R.O. 1978. Vegetational patterns on the Tittabawassee floodplain at the Goetz Grove Nature Center, Saginaw, Michigan. *Michigan Botanist* **17**: 91–9.

Whitlow, T.H. and Harris, R. 1979. Flood tolerance in plants: a state-of-the-art review. US Army Engineers Waterways Experiment Station Technical Report E-79-2.

Whittaker, R.H. 1967. Gradient analysis of vegetation. *Biological Reviews* **42**: 20–64.

Whittaker, R.H. and Levin, S.A. 1977. The role of mosaic phenomena in natural communities. *Theoretical Population Biology* **12**: 117–39.

Wikum, D.A. and Wali, M.K. 1974. Analysis of a North Dakota gallery forest: vegetation in relation to topographic and soil gradient. *Ecological Monographs* **44**: 441–64.

Wilcock, D.N. and Essery, C.I. 1991. Environmental impacts of channelization on the river Main, County Antrim, Northern Ireland. *Journal of Environmental Management* **32**: 127–43.

Wildi, O. 1989. Analysis of the disintegrating group and gradient structure in Swiss riparian forests. *Vegetatio* **83**: 179–86.

Wilkin, D.C. and Hebel, S.J. 1982. Erosion, redeposition, and delivery of sediment to midwestern streams. *Water Resources Research* **18**: 1278–82.

Willgoose, G., Bras, R.L. and Rodriguez-Iturbe, I. 1991*a*. Results from a new model of river basin evolution. *Earth Surface Processes and Landforms* **16**: 237–54.

Willgoose, G., Bras, R.L. and Rodriguez-Iturbe, I. 1991*b*. A coupled channel network growth and hillslope evolution model. 1. Theory. *Water Resources Research* **27**: 1671–84.

Willgoose, G., Bras, R.L. and Rodriguez-Iturbe, I. 1991*c*. A coupled channel network growth and hillslope evolution model. 2. Nondimensionalization and applications. *Water Resources Research* **27**: 1685–96.

Williams, D.D. 1987. *The Ecology of Temporary Waters*. Timber Press, Portland.

Williams, G.P. 1978. Historical perspective of the Platte Rivers in Nebraska and Colorado. In W.D. Graul and S.J. Bissell (eds.) *Lowland River and Stream Habitat in Colorado: A Symposium*, pp. 11–41. Colorado Division of Wildlife, Denver.

Williams, G.P. and Wolman, M.G. 1984. Downstream effects of dams on alluvial rivers. US Geological Survey Professional Paper 1286.

Williams, J.E., Kobetich, G.C. and Benz, C.T. 1984. Management aspects of relict populations inhabiting the Amargosa Canyon ecosystem. In R.E. Warner and K.M. Hendrix (eds.) *California Riparian Systems*, pp. 706–15. University of California Press, Berkeley.

Williams, J.G. 1989. Interpreting physiological data from riparian vegetation: cautions and complications. In D.L. Abell (ed.) Proceedings of the California Riparian Systems Conference. US Forest Service General Technical Report PSW-110, pp. 381–6.

Williams-Linera, G. 1990. Vegetation structure and environmental conditions of forest edges in Panama. *Journal of Ecology* **78**: 356–73.

Wilson, R.E. 1970. Succession in stands of *Populus deltoides* along the Missouri River in southeastern South Dakota. *American Midland Naturalist* **83**: 330–42.

Wischmeier, W.H. and Smith, D.D. 1965. *Predicting rainfall erosion losses from cropland east of the Rocky Mountains*. Agriculture Handbook 282, US Department of Agriculture, Washington.

Wissmar, R.C. and Swanson, F.J. 1990. Landscape disturbances and lotic ecotones. In R.J. Naiman and H. Decamps (eds.) *The Ecology and Management of Aquatic-Terrestrial Ecotones*, pp. 65–89. Man and the Biosphere Series Vol. 4. UNESCO, Paris.

Wistendahl, W.A. 1958. The flood plain of the Raritan River, New Jersey. *Ecological Monographs* **28**: 129–53.

Witinok, P.M. 1991. A critical examination of the Universal Soil Loss Equation (USLE). *Geographical Perspectives* **62**: 25–54.

Wittwer, R.F., Immel, M.J. and Ellingsworth, F.R. 1980. Nutrient uptake in fertilized plantations in American sycamore. *Soil Science Society of America Journal* **44**: 606–10.

Woldenberg, M. and Horsfield, K. 1983. Finding the optimal lengths for three branches of a junction. *Journal of Theoretical Biology* **104**: 301–18.

Wolman, M.G. 1959. Factors influencing erosion on a cohesive river bank. *American Journal of Science* **257**: 204–16.

Wolman, M.G. and Leopold, L.B. 1957. River flood plains: some observations on their formation. US Geological Survey Professional Paper 282-C.

Woodward-Clyde Associates 1980. Gravel removal studies in Arctic and Subarctic floodplains in Alaska. US Fish and Wildlife Service OBS-80/08.

Worster, D. 1985. *Rivers of Empire*. Pantheon Books, New York.

Wyant, J.G. and Ellis, J.E. 1990. Compositional patterns of riparian woodlands in the Rift Valley of northern Kenya. *Vegetatio* **89**: 23–37.

Yanosky, T.M. 1982*a*. Effects of flooding upon woody vegetation along parts of the Potomac River flood plain. US Geological Survey Professional Paper 1206.

Yanosky, T.M. 1982*b*. Hydrologic inferences from ring widths of flood-damaged trees, Potomac River, Maryland. *Environmental Geology* **4**: 43–52.

Yanosky, T.M. 1983. Evidence of floods on the Potomac River from anatomical abnormalities in the wood of flood-plain trees. US Geological Survey Professional Paper 1296.

Yeager, L.E. 1949. Effects of permanent flooding in a river-bottom timber area. *Illinois Natural History Survey Bulletin* **25**: 33–65.

Yon, D. and Tendron, G. 1981. *Alluvial forests of Europe*. Nature and Environment Series No. 22. Council of Europe, Strasbourg.

Young, C.E., Klawitter, R.A. and Henderson, J.E. 1972. Hydrologic model of a wetland forest. *Journal of Soil and Water Conservation* **27**: 122–4.

Youngblood, A.P. and Zasada, J.C. 1991. White spruce regeneration options on river floodplains in interior Alaska. *Canadian Journal of Forest Research* **21**: 423–33.

Zajonc, J. 1985. Earthworm (*Lumbricidae*) community. In M. Penka, M. Vyskot, E. Klimo, and F. Vasicek (eds.) *Floodplain Forest Ecosystem*. I. *Before Water Management Measures*, pp. 373–85. Elsevier, Amsterdam.

Zejda, J. 1985. Energy flow through the small mammal community of a floodplain forest. In M. Penka, M. Vyskot, E. Klimo, and F. Vasicek (eds.) *Floodplain Forest Ecosystem*. I. *Before Water Management Measures*, pp. 357–71. Elsevier, Amsterdam.

Zidek, V. 1988. Actual and potential evapotranspiration in the floodplain forest. *Ekologia (CSSR)* **7**: 43–59.

Zipp, L.S. 1988. Tree seedling survival in a southeastern Iowa floodplain forest. Unpublished research paper, Department of Geography, University of Iowa.

Zimmerman, R.C. 1969. Plant ecology of an arid basin: Tres Alamos – Reddington area, southeastern Arizona. US Geological Survey Professional Paper 485D: 1–51.

Zinke, A. and Gutzweiler, K.-A. 1990. Possibilities for regeneration of floodplain forests within the framework of the flood-protection measures on the Upper Rhine, West Germany. *Forest Ecology and Management* **33/34**: 13–20.

Zonneveld, I.S. 1990. Scope and concepts of landscape ecology as an emerging science. In I.S. Zonneveld and R.T.T. Forman (eds.) *Changing Landscapes: An Ecological Perspective*, pp. 3–20. Springer-Verlag, New York.

Zube, E.H. and Simcox, D.E. 1987. Arid lands, riparian landscapes, and management conflicts. *Environmental Management* **11**: 529–35.

Index